0

NONLINEAR ORDINARY DIFFERENTIAL EQUATIONS IN TRANSPORT PROCESSES

This is Volume 42 in
MATHEMATICS IN SCIENCE AND ENGINEERING
A Series of Monographs and Textbooks
Edited by RICHARD BELLMAN, *University of Southern California*

The complete listing of books in this series is available from the Publisher upon request.

NONLINEAR ORDINARY DIFFERENTIAL EQUATIONS IN TRANSPORT PROCESSES

WILLIAM F. AMES

DEPARTMENT OF MECHANICS AND HYDRAULICS
UNIVERSITY OF IOWA
IOWA CITY, IOWA

1968

ACADEMIC PRESS   New York San Francisco London
A Subsidiary of Harcourt Brace Jovanovich, Publishers

210288

ACADEMIC PRESS INC.
111 Fifth Avenue, New York, New York 10003

United Kingdom Edition published by
ACADEMIC PRESS, INC. (LONDON) LTD.
24/28 Oval Road, London NW1

LIBRARY OF CONGRESS CATALOG CARD NUMBER: 68-18651

PRINTED IN THE UNITED STATES OF AMERICA
79 80 81 82   9 8 7 6 5 4 3

*For stimulating a love of scholarship
and for understanding
I dedicate this volume
to my parents and my wife.*

PREFACE

The characteristics of transport phenomena are essentially nonlinear—a fact that has long been recognized. Indeed the advances in the analytic theory of fluid mechanics have rested heavily upon and have contributed greatly to our knowledge of nonlinear differential equations, both ordinary and partial. Thus it is surprising to note that no bound volume exists which unifies a number of available methods for solving nonlinear ordinary differential equations in transport problems. On the other hand books on nonlinear mechanics abound.

This volume is primarily concerned with methods of solution for nonlinear ordinary differential equations arising from transport processes—which I herein interpret to mean chemical kinetics, fluid mechanics, diffusion, heat transfer, and related areas. Examples will usually originate in these areas. The book does not reproduce, in yet another form, the already extensive works in vibrations, electronics, stability, etc. that constitute the general area of nonlinear mechanics.

The exposition is a mixture of theory and examples. A chapter on the origin of some equations is followed by chapters on exact methods, detailed examples, approximate methods, and numerical methods.

In this rapidly advancing field a volume without a substantial bibliography would be of limited use. An effort has been made to supply a useful current bibliography. Thus the source for amplification of the text material is readily found. Over 280 references, some to volumes themselves heavily referenced, are included.

I am indebted to all the researchers in this subject. For encouragement and criticism I thank my colleagues and friends J. R. Ferron, S. E. Jones, and J. R. Vinson.

WILLIAM F. AMES

September 1967
Iowa City, Iowa

CONTENTS

xi

Chapter 3. Examples from Transport Phenomena

Chapter 4. Approximate Methods

Chapter 5. Numerical Methods

Appendix. Similarity Variables by Transformation Groups

1

THE ORIGIN OF NONLINEAR EQUATIONS

Introduction

Why the intense interest in the nonlinear problems of science and technology? Perhaps the best reason is that essentially all of those problems *are nonlinear* from the outset. The linearizations commonly employed are an approximating device arising from our confession that we are incapable of facing the challenge presented by the nonlinear problems as such. However, we hasten to add that such linearizations, as approximating devices are and always will be valuable. In many instances they suffice to provide all the information we need. In others they provide gross approximations and direction. In still a third group we find that solutions to the nonlinear problem are obtainable by means of a sequence of linear problems. The linearization allows us to bring to bear on the problem that considerable domain of mathematics which has been structured on the fundamental postulate of linearity. Linear mathematics has been extensively "mined."

On the other hand, there are many cases in which linearization and the subsequent treatment of that system is not sufficient. Essentially new phenomena frequently occur in nonlinear problems which cannot occur in linear systems. In nonlinear vibrations there exists a surprisingly wide variety of such new phenomena. Examples include the *jump in amplitude* as frequency varies, the occurrence of *subharmonic* forced oscillations, the occurrence of *combination* tones and the occurrence of *relaxation* phenomena. In gas dynamics an essentially nonlinear phenomenon is the formation of a *discontinuous shock wave* from a smooth wave. Surprisingly,

1

these phenomena can be investigated by procedures which are interesting and reasonably elementary. In a large measure they do not employ sophisticated mathematics. Indeed the great stimuli in this area, and hence many of the methods, have been generated by technologists. Many, when faced with these problems admit defeat. A few face the challenge. It is to these few that we are indebted.

1.1 What Is Nonlinearity?

In many engineering investigations perhaps the most common procedure is to mention nonlinearities merely to dismiss them. In books and journals alike we find such phrases as "given a perfect insulator," "let the modulus be constant," "consider small amplitude vibration," "for constant thermal conductivity," "let the fluid be inviscid," and so on. In elementary college courses we often get the impression that everything is ideal, frictionless, inelastic, rigid, inviscid, incompressible, and the like.

What in fact is the complication that causes us to mutter these phrases? Recall that the derivative, denoted by D, possesses the fundamental property "the derivative of the sum of two functions f and g is equal to the sum of the derivative of the functions": that is for any constants a, b,

$$D[af + bg] = a\,Df + b\,Dg. \tag{1.1}$$

This property is also possessed by the integral, all difference operators, and combinations of the above classes. In short this linearity property is possessed by the basic operators used to construct our fundamental mathematical tools, the differential equations, integral equations, difference equations, and so forth. However, the mere possession of linearity by the operators *does not ensure* that the complete equation will have this property.

In general we say that an operator L is linear if

$$L[af + bg] = aLf + bLg \tag{1.2}$$

where f, g are functions and a, b are constants. It is clear that the logarithm operator, squaring operator, exponential operator are examples which do not possess this property. Indeed recall that

$$\exp[af + bg] = \exp[af]\,\exp[bg]$$
$$\neq a\exp f + b\exp g. \tag{1.3}$$

Consider the linear ordinary differential equation

$$\frac{d^2x}{dt^2} + ax = 0 \qquad (1.4)$$

where a is constant. If x_1 and x_2 are both solutions of Eq. (1.4), i.e.,

$$\frac{d^2x_1}{dt^2} + ax_1 = 0, \qquad \frac{d^2x_2}{dt^2} + ax_2 = 0$$

then it follows, by the linearity of the second derivative that $Ax_1 + Bx_2$ is a solution, where A and B are arbitrary constants. This fact is the foundation of the *principle of superposition* which has been essentially responsible for the great successes of the past in constructing effective theories for (linearized) physical phenomena. In accordance with this principle elementary solutions of the pertinent mathematical equation could be combined to yield more flexible ones, namely ones which could satisfy the auxiliary conditions describing the particular physical phenomena.

Let us now suppose that Eq. (1.4) is modified to account for the nonlinear restoring force $ax + bx^2$. The equation now becomes

$$\frac{d^2x}{dt^2} + ax + bx^2 = 0. \qquad (1.5)$$

If two solutions $x_1(t)$ and $x_2(t)$ of Eq. (1.5) have been found, is $x_1 + x_2$ also a solution? Upon substituting $x_1 + x_2$ for x we find

$$\frac{d^2x_1}{dt^2} + \frac{d^2x_2}{dt^2} + ax_1 + ax_2 + bx_1{}^2 + 2bx_1x_2 + bx_2{}^2$$

which may be grouped as

$$\left[\frac{d^2x_1}{dt^2} + ax_1 + bx_1{}^2\right] + \left[\frac{d^2x_2}{dt^2} + ax_2 + bx_2{}^2\right] + 2bx_1x_2. \qquad (1.6)$$

The first two bracketed expressions of relation (1.6) vanish by assumption but the last term is not zero. Hence $x_1 + x_2$ is not a solution. *Thus the principle of superposition no longer holds and we have suffered a major setback.* It is the loss of this principle and the lack of an effective replacement that leads to many difficulties. On occasion we can perform some transformation that reduces the nonlinear problem to a linear one and therefore utilize the principle of superposition.

By this time the reader should be fully aware that the use of the phrases linear and nonlinear does not refer to the graphical character of a function. Rather they concern the properties of operators.

If in an ordinary differential equation the dependent variable y and its derivatives are of the *first degree* only and no product of these terms, such as $y'y'''$, $yy'y''$, occur then the equation is clearly linear. Thus $y'' + x^3 y + e^x y = \log x$ is a linear equation in y. The presence of x^3, e^x, and $\log x$, or any other function of the independent variable x, does not constitute nonlinearity in y. But $y''' + y'y'' = 0$, $(y'')^2 + y' + y^{1/2} = 0$ are nonlinear equations of the third and second order, respectively. The nonlinear terms are $y'y''$, $(y'')^2$, and $y^{1/2}$. An immediate observation is that the types of nonlinearities are legion. Some are more difficult to handle than others. Such a great variety leads one to suspect that no single theory of nonlinearity is possible. Beginnings have been made for certain classes but the wasteland is poorly charted and full of "quagmires."

1.2 Other Departures from Linear Theory

A *linear* differential equation of the nth order has n distinct or linearly independent solutions. Suppose that y_1 and y_2 are two linearly independent (i.e., not proportional) solutions of

$$y'' + b(x)y' + c(x)y = 0. \tag{1.7a}$$

The general solution of Eq. (1.7a) is

$$y = Ay_1 + By_2 \tag{1.7b}$$

where A and B are arbitrary constants. If Eq. (1.7a) is modified by the addition of the term yy', it becomes nonlinear, and reads

$$y'' + (b(x) + y)y' + c(x)y = 0. \tag{1.8}$$

If y_1 and y_2 are solutions of Eq. (1.8) their sum is not a solution. Indeed the concept of linear independence is meaningless—there cannot be linearly independent solutions of a nonlinear equation. Thus we must *discard* this concept.

Secondly, Eq. (1.8) may be considered to have a general solution, obtainable by two integrations, involving two arbitrary constants. The general solution is a *function* of these constants and not a simple form analogous to Eq. (1.7b). As an illustration of this concept we consider an example which occurs in a boundary layer problem in aeronautics. The equation describing the process of fluid separation from an airfoil surface is

$$y \frac{d^2 y}{d\eta^2} + v \left(\frac{dy}{d\eta}\right)^2 = 0 \tag{1.9}$$

where $v = (1 - \mu)/\mu$. Setting $p = dy/d\eta$, we find

$$\frac{d^2 y}{d\eta^2} = \frac{d}{d\eta}\left(\frac{dy}{d\eta}\right) = \frac{d}{dy}(p)\frac{dy}{d\eta} = p \frac{dp}{dy}, \tag{1.10}$$

so that Eq. (1.9) reduces to the first order equation

$$yp \frac{dp}{dy} = -vp^2. \tag{1.11}$$

This equation may be rewritten in separable form as

$$\frac{dp}{p} = -v \frac{dy}{y}$$

whose solution, upon integration is

$$\log p = -v \log y + C_1$$

or

$$p = \frac{dy}{d\eta} = Cy^{-v} = Cy^{1-(1/\mu)}. \tag{1.12}$$

Integrating again we find

$$y^{1/\mu} = \frac{1}{\mu}(Cx + B)$$

or

$$y = \left[\frac{1}{\mu}(Cx + B)\right]^\mu, \tag{1.13}$$

wherein C and B are the two constants of integration. *The general solution is a function of these constants.*

Third, unless all terms of the nonlinear equation are of the same degree in y, a constant A ($\neq 1$) times a solution is not a solution. Consider the Van der Pol equation

$$y'' + \varepsilon y^2 y' - \varepsilon y' + y = 0. \tag{1.14}$$

The first, third, and fourth terms are of degree one in y and its derivatives while the second term is of degree three. If y_1 is a solution of Eq. (1.14) Ay_1 ($A \neq 1$, $A \neq 0$) is not a solution, for upon substitution we have the relation

$$Ay_1'' + \varepsilon A^3 y_1^2 y_1' - \varepsilon A y_1' + A y_1.$$

Upon division by A this relation becomes

$$y_1'' + \varepsilon A^2 y_1^2 y_1' - \varepsilon y_1' + y_1 \neq 0 \tag{1.15}$$

unless $A = 1$.

On the other hand if all terms are of the same degree, such as those of Eq. (1.9), then a constant times a solution is also a solution. One easily verifies that $A[(1/\mu)(Cx + B)]^\mu$ is also a solution of Eq. (1.9) for any constant A.

The above results indicate that we must drastically modify our thinking when nonlinear problems are attempted. Our indoctrination into the methods of linear analysis has been long and extensive. It is not easy to reject a considerable portion of this orientation. But reject it we must.

1.3 Literature

Refinements and advances in science and technology, the demands of the space age and the availability of computing machines has stimulated great interest in a whole range of nonlinear problems. Much of this work originated in astronomy and nonlinear mechanics. In astronomy one encounters the Lane–Emden equation for the gravitational equilibrium of gaseous configurations. It originated with Lane [1] (1870) and has been further investigated by Emden [2] and Kelvin [3]. A more recent treatment is given by Chandrasekhar [4] and in more general form by Davis [5].

Mechanical and electrical oscillations have concerned numerous researchers. Here we find the works of Andronow and Chaikin [6], Stoker

[7], McLachlan [8], Kryloff and Bogoliuboff [9], Minorsky [10–12] Cunningham [13], and Malkin [14]. Questions of stability and other theoretical concepts are explored in the monographs by Hale [15], Lasalle and Lefschetz [16, 17] and Lefschetz [18].

The theory of elasticity and plasticity is rich in nonlinear equations. Probably the earliest problem was that of the "elastica" discussed by McLachlan [8]. Bickley [19] has modified the idea to characterize the drape of fabrics by means of a "bending length." A substantial number of pre-1940 papers are summarized by von Karman [20]. This excellent summary concerns the large deflection of elastic structures as well as problems in plasticity and fluid mechanics. It is partially concerned with nonlinear partial differential equations (see also Ames [21])—we shall see that this area and our present one are intimately related.

Extensive research in connection with Newtonian and non-Newtonian fluids has been conducted by many authors. Here we mention only the work of the pioneers Prandtl, Blasius, Goldstein, Bickley, Karman, Oseen, Reynolds, and Taylor (see Karman [20] and Schlichting [22]). Considerable attention will be paid to problems arising in fluid mechanics and related areas in this volume.

In electricity nonlinearity arises from saturation of ironcored apparatus. The triode oscillator is an example of a nonlinear device. Van der Pol and Appleton were leaders in triode problems and the Van der Pol equation (1.14) has served as a fundamental model for many investigations. Under certain conditions the triode can undergo relaxation oscillations, a phenomenon which occurs in physiology (the heart beat), elasticity (flag waving), and in the hunting action of motor generator equipment (see Cunningham [13]).

Lastly we would be neglectful in not including the work of Volterra [23] on the growth and competition among biological species. Special cases are in fact characterized by relaxation oscillations. Further work in the biological field is given by Rashevsky [24].

In addition to the above references two other books have appeared which bear mention because of their different orientation. These are the work by Struble [25] and by Saaty and Bram [26]. Both books aim at the development of embryo theories as indeed do other studies (see Hale [15], Lasalle and Lefschetz [17, 18]).

The plan of this monograph is to emphasize those fields that have essentially been neglected in book form. It will not reproduce, in yet another form, the already extensive works in vibrations, electronics, stability, etc. that constitute the general area of nonlinear mechanics. Thus we concentrate our attention on those nonlinear differential equations which arise in chemical kinetics, fluid mechanics, diffusion and heat transfer, and related areas. Our examples will usually originate from this area. Often we shall appeal to the author's companion volume [21] for concepts from that subject.

1.4 Examples in Kinetics

Whatever the postulate utilized the equations for complex chemical reactions are nonlinear. A discussion of the foundations is given by Benson [27]. Let us consider the irreversible reaction

$$aA + bB \rightarrow dD + eE \qquad (1.16)$$

where a molecules of species A react with b molecules of B to form the products. We take as our fundamental postulate the law of "mass action"— that is, in dilute solutions, the speed of a chemical reaction is proportional to the molecular concentration of the substances reacting. We formulate the problem following the rate of disappearance of A. Let $x_A = x_A(t)$ represent the number of molecules of A present at any time t. Then dx_A/dt represents the rate of disappearance of A and therefore is negative. Clearly dx_A/dt is proportional to x_A for if x_A is doubled twice as many molecules are available for reaction, resulting in double the number of molecules disappearing per unit time. If we rewrite Eq. (1.16) as

$$1 \cdot A + (a - 1)A + bB \rightarrow dD + eE \qquad (1.17)$$

attention can be focused upon one molecule of A. For the reaction to occur this molecule must "collide" simultaneously with $a - 1$ molecules of A and b molecules of B. Hence, the rate of disappearance of A is proportional to the number of simultaneous interactions of the type just mentioned. The number of these interactions is proportional to the product of the concentrations of the reacting molecules. The concentration (denoted C_A, C_B, etc.) of each species is raised to a power corresponding to the number of molecules of that species which must interact with the molecule of A considered.

For constant temperature the equation expressing the rate of disappearance of A is

$$\frac{dx_A}{dt} = -kx_A\, C_A^{a-1}\, C_B^{\,b} \qquad (1.18)$$

where k is a reaction rate constant. If the reaction temperature changes, k being a function of temperature varies. This variation has been extensively investigated and the usual form employed is that of Arrhenius (see Benson [27])

$$k = k_0 \exp[-\beta/T] \qquad (1.19)$$

where T is the absolute temperature and k_0, β are reaction constants.

In the special case of a reaction at constant volume V, we have $C_A = x_A/V$ so that Eq. (1.18) becomes

$$\frac{dC_A}{dt} = -k\, C_A^{\,a}\, C_B^{\,b}. \qquad (1.20)$$

A multiplicity of these equations occur if a number of simultaneous reactions are involved. Several complex reactions are discussed by Ames [28, 29] and will be investigated later. For the present we list only the equations for the constant volume gas phase pyrolysis of toluene (see Benson [27]) given by the reactions

$$A_1 \xrightarrow{k_1} A_2 + A_3$$

$$A_3 + A_1 \xrightarrow{k_2} A_2 + A_4$$

$$A_3 + A_1 \xrightarrow{k_3} A_5 + A_6 \qquad (1.21)$$

$$A_6 + A_1 \xrightarrow{k_4} A_2 + A_7$$

$$A_2 + A_2 \xrightarrow{k_5} A_3.$$

Setting x_i = concentration of A_i the kinetic equations are

$$\frac{dx_1}{dt} = -k_1 x_1 - k_2 x_1 x_2 - k_3 x_1 x_3 - k_4 x_1 x_6$$

$$\frac{dx_2}{dt} = k_1 x_1 + k_2 x_1 x_3 + k_4 x_1 x_6 - k_5 x_2^{\,2}$$

$$\frac{dx_3}{dt} = k_1 x_1 - k_2 x_1 x_3 - k_3 x_1 x_3 + k_5 x_2{}^2$$

$$\frac{dx_4}{dt} = k_2 x_1 x_3 \tag{1.22}$$

$$\frac{dx_5}{dt} = k_3 x_1 x_3$$

$$\frac{dx_6}{dt} = k_3 x_1 x_3 - k_4 x_1 x_6$$

$$\frac{dx_7}{dt} = k_4 x_1 x_6 .$$

One of the problems involved in this system is the determination of the rate constants k_i from data.

1.5 Heat Transfer and Chemical Reaction

If Q is the monomolecular heat of reaction, k the thermal conductivity and W the reaction velocity then the equation specifying the thermal balance between the heat generated by the exothermic chemical reaction and that conducted away is

$$k \, \nabla^2 T = -QW. \tag{1.23}$$

If we use the Arrhenius relation for W then

$$W = C\lambda \exp[-E/RT] \tag{1.24}$$

where T, C, λ, E, R, and ∇^2 are respectively temperature, concentration, frequency factor, activation energy, gas constant, and Laplace operator, respectively. For T sufficiently close to T_0 we write

$$-\frac{E}{RT} = -\frac{E}{RT_0}\left[\frac{1}{1 + (T - T_0)/T_0}\right] \approx -\frac{E}{RT_0}\left[1 - \frac{T - T_0}{T_0}\right]. \tag{1.25}$$

Setting

$$\theta = (E/RT_0{}^2)(T - T_0)$$

$$\delta = (QE/kRT_0{}^2)C\lambda \exp(-E/RT_0) \tag{1.26}$$

we find the equation for θ to be

$$\nabla^2\theta + \delta \exp \theta = 0. \tag{1.27}$$

In one dimension Eq. (1.27) becomes

$$\frac{d^2\theta}{dZ^2} + \frac{n}{Z}\frac{d\theta}{dZ} = -\delta \exp \theta \tag{1.28}$$

where $n = 0$ gives the equation in rectangular coordinates, $n = 1$ circular cylindrical coordinates, and $n = 2$ spherical coordinates.

Equations of the form (1.28) also occur in the theory of space charge about a glowing wire as described by Richardson [30], in the nebular theory for mass distribution of gaseous interstellar material under the influence of its own gravity (see Walker [31]) and in vortex motion of incompressible fluids (Bateman [32]). The cases $n = 0$, 1 have been solved exactly as we shall see later. The paper of Chambré [33] is the basic one here.

1.6 Equations from ad hoc Methods for Partial Differential Equations

The ad hoc methods (see Ames [34]) that have been constructed to obtain particular solutions of nonlinear partial differential equations generate a potpourri of nonlinear ordinary differential equations. As in the linear theory this is one of the greatest sources for the equations which concern us here.

A. TRAVELING WAVE SOLUTIONS

Burgers [35], Hopf [36], and Cole [37] have considered the Burgers' equation

$$u_t + uu_x = vu_{xx} \tag{1.29}$$

first introduced as a mathematical model of turbulent flow. Setting $u = \psi_x$, integrating with respect to x and discarding an arbitrary function of time, we have

$$\psi_t + \tfrac{1}{2}\psi_x{}^2 = v\psi_{xx}. \tag{1.30}$$

We now attempt an ad hoc solution of the *generalized traveling wave type*

$$\dot{\psi} = f[t + g(x)] \tag{1.31}$$

where f and g are to be determined. Upon substitution into Eq. (1.30) we obtain, with $w = t + g(x)$,

$$f'(w) [1 - vg''(x)] = [g'(x)]^2 [vf'' - \tfrac{1}{2}(f')^2] \tag{1.32}$$

which may be rewritten as

$$[1 - vg''(x)]/(g')^2 = [vf'' - \tfrac{1}{2}(f')^2]/f' = \lambda \tag{1.33}$$

where λ is a constant. λ must be constant for the left hand side of Eq. (1.33) depends only upon x and the right hand side on both t and x. This is the classical separation argument. Thus $f(w)$ and $g(x)$ must satisfy the equations

$$\begin{aligned} vf'' - \tfrac{1}{2}(f')^2 - \lambda f' &= 0 \\ vg'' + \lambda(g')^2 &= 1; \end{aligned} \tag{1.34}$$

clearly they are a nonlinear set.

B. SEPARATION OF VARIABLES

Various researchers have attempted a direct separation of specific nonlinear partial differential equations. Among these we find the work of Oplinger [38, 39] in the nonlinear vibration of threads, Smith [40] in anisentropic flow, Tomotika and Tamada [41] in two dimensional transonic flow and Keller [42]. All these authors use the *modus operandi* of examining the form of the solutions obtainable by this process and then investigate what auxiliary conditions can be satisfied. Since superposition is no longer available it is trite to remark that not all auxiliary conditions are compatible with these solutions.

Oplinger's problem is that of solving the "weakly" nonlinear hyperbolic equation

$$y_{tt} - C^2\left[1 + \alpha \int_0^1 y_x{}^2 \, dx\right] y_{xx} = 0 \tag{1.35}$$

governing the oscillations of a string fixed at one end $(x = 0)$ and subject to an unspecified periodic oscillation at the other end $(x = 1)$.

Upon attempting direct separation with $y = F(x)G(t)$ it is evident that $\int_0^1 y_x^2 \, dx = G^2(t) \int_0^1 (dF/dx)^2 \, dx = IG^2(t)$ so then

$$F'' + v^2 F = 0$$
$$G'' + v^2 C^2 [1 + \alpha IG^2] G = 0. \tag{1.36}$$

These separation equations involve the constant v. Clearly only the second is nonlinear.

As a second example we note that Birkhoff [43] introduced the development of similarity variables by means of separation of variables. For the general concept we refer to Ames [21] or Hansen [44]. Let us consider the nonlinear diffusion equation (with $D = C^n$)

$$[C^n C_r]_r = C_t. \tag{1.37}$$

The general concept is to search for new variables (ξ, η),

$$\xi = \xi(r, t), \qquad \eta = \eta(r, t) \tag{1.38}$$

such that when Eq. (1.37) is transformed to these new variables it becomes separable. For simplicity we demonstrate with

$$\xi = t, \qquad \eta = r/R(t) \tag{1.39}$$

where $R(t)$ is to be determined so that the equation is separable—that is Eq. (1.37) has the separated solution

$$C = U(t)\, Y(\eta). \tag{1.40}$$

Upon substitution of Eq. (1.40) into Eq. (1.37) and division by $U^{n+1}(t)$ we find

$$\frac{d}{d\eta}[Y^n Y'] = [R(t)/U^{n+1}][RU'Y - \eta UR'Y']. \tag{1.41}$$

The left hand side of Eq. (1.41) is a function only of η. Thus the solution for C is separable if the term in the square bracket on the right is a function of t times a function of η. This occurs nontrivially if and only if

$$UR' = -AU'R \tag{1.42}$$

whereupon

$$R(t) = [U(t)]^{-A} \tag{1.43}$$

with A constant. Eq. (1.41) now takes the form

$$\frac{\frac{d}{d\eta}[Y''Y']}{Y + A\eta Y'} = \frac{U'}{U^{n+1+2A}} = -\lambda \tag{1.44}$$

with λ constant. Two nonlinear equations must now be solved for U and Y. The values of λ and A depend upon boundary conditions.

C. ELEMENTARY EXPANSION

Two or three term expansion solutions are sometimes possible for nonlinear partial differential equations. To describe the development of one such expansion we utilize the (hypothetical gas) transonic equation of Tomotika and Tamada [41]

$$[kw]_{\psi\psi} = [(kw)^2]_{\Phi\Phi} . \tag{1.45}$$

The authors found that Eq. (1.45) possessed solutions of the form

$$kw = f(\Phi) + g(\psi) \tag{1.46a}$$

$$kw = f_0(\psi) + f_1(\psi)\Phi^2 \tag{1.46b}$$

$$kw = F(\Phi + \psi^2) + 2\psi^2 \tag{1.46c}$$

as well as a traveling wave and separate solution. In all cases nonlinear ordinary differential equations are generated. For example, upon substitution of Eq. (1.46b) into Eq. (1.45) we find

$$f_0'' - 4f_1 f_0 = [12f_1{}^2 - f_1'']\Phi^2. \tag{1.47}$$

The left hand side of Eq. (1.47) depends only upon ψ, while the right involves both variables. This is clearly impossible unless

$$f_1'' - 12f_1{}^2 = 0, \tag{1.48}$$

that is, a nonlinear equation for f_1.

D. FUNCTIONAL METHODS

Cole [37] introduced an ingenious idea in transforming the modified Burgers' equation (1.30) into the linear diffusion equation $\theta_t = \nu\theta_{xx}$. We begin with Eq. (1.30) and set

$$\psi = F[\theta(x, t)] \tag{1.49}$$

where both F and θ may be selected to simplify the equation. Under this transformation on ψ we find Eq. (1.30) becomes

$$F'[\theta_t - v\theta_{xx}] = [vF'' - \tfrac{1}{2}(F')^2]\theta_x^2 \tag{1.50}$$

where the prime indicates differentiation with respect to θ. If we choose θ as a solution of $\theta_t = v\theta_{xx}$ then $\theta_x \neq 0$, so F must be a solution of the non-linear equation

$$2vF'' = (F')^2. \tag{1.51}$$

E. Equation Splitting

The success of the *equation splitting* idea of Ames [34, Eqs. (76)–(78)] depends strongly upon our ability to solve the equations which concern us here. This procedure, now (1966) in its infancy generates a potpourri of equations. In Ames [34] we meet the equation

$$A\alpha^2\eta'' = \alpha^2(\eta')^2 + 1. \tag{1.52}$$

F. Relations between Dependent Variables

Insight into the solution character of some partial differential equations may be obtained by assuming an unspecified relation between the dependent variables. The resulting simplification can be striking.

Consider the boundary layer flow of a non-Newtonian ("power law") fluid over a semi-infinite flat plate,

$$uu_x + vu_y = v[(u_y)^n]_y$$
$$u_x + v_y = 0. \tag{1.53}$$

We shall investigate forms of the solution for which

$$v = F(u) \tag{1.54}$$

for general $F(u)$. From the continuity equation we find $u_x = -F'(u)u_y$ so that the momentum equation becomes

$$[uF'(u) - F]u' + v[(u')^n]' = 0, \tag{1.55}$$

that is, an ordinary differential equation in y, with parameter x. Selecting $F = \alpha u^{m-1}$, $m > 1$ we find that Eq. (1.55) becomes

$$v[(u')^n]' + \alpha(m - 2)u^{m-1} u' = 0 \qquad (1.56)$$

which integrates once to

$$v(u')^n + (\alpha/m)(m - 2)u^m = A(x) \qquad (1.57)$$

for $m \neq 2$ and $A(x)$ is to be determined from the auxiliary conditions.

1.7 Equations from Similarity Solutions

Many advances in fluid mechanics, nonlinear diffusion, wave propagation and other fields have resulted from our ability to construct similarity variables. Of the four methods commonly used (see Hansen [44]) the group invariant has the soundest mathematical background (see Ames [21]). It is easily generalized and is therefore preferred by the present author. In this section we summarize some nonlinear equations obtained by Lee and Ames [45] in investigating heat transfer in power law fluids.

From Lee and Ames [45] the basic equations in boundary layer form are

$$u_x + v_y = 0 \qquad (1.58)$$

$$uu_x + vu_y = U_e U_e' + [\,|u_y|^{n-1}u_y]_y + \alpha\theta \qquad (1.59)$$

$$u\theta_x + v\theta_y = (N_{Pr})^{-1}\,\theta_{yy} \qquad (1.60)$$

where u, v, U_e, N_{Pr}, θ, and n represent dimensionless velocities in the dimensionless x and y directions, velocity outside the boundary layer, Prandtl number, dimensionless temperature, and power law parameter, respectively. For forced convection $\alpha = 0$ and for natural convection $\alpha = 1$.

(a) *Momentum Transfer (Only) over a Flat Plate*

$$n(n + 1)f''' + (f'')^{2-n} f = 0. \qquad (1.61)$$

Research in this problem has also been carried out by Acrivos *et al.* [46].

(b) *Momentum Transfer (Only) in Jet Flow*

$$(f')^2 + ff'' = \frac{d}{d\eta}\,[\,|f''|^{n-1} f'']. \qquad (1.62)$$

Gutfinger and Shinnar [47], Kapur [48] and Kapur and Srivastava [49] have extensively studied the engineering implications of this kind of flow.

(c) *Falkner–Skan Flows* $U_e = x^m$

$$(f')^2 - \frac{(2n-1)m+1}{m(n+1)} ff'' = 1 + \frac{1}{m}\frac{d}{d\eta}[(f'')^n].\tag{1.63}$$

(d) *Wedge Flow, Forced Convection* $U_e = x^{1/3}$, $\theta_B = x^t$ (1.64)

$$f'^2 - 2ff'' = 1 + 3\frac{d}{d\eta}[(f'')^n],$$

$$g'' + N_{Pr}(\tfrac{2}{3}fg' - tf'g) = 0.$$

(1.65)

(e) *Forced Convection, Variable Thermal Conductivity*

$$n(n+1)f''' + (f'')^{2-n}f = 0,$$

$$\frac{1-n}{(r-1)(n+1)}f'g - \frac{1}{n+1}fg' = \frac{1}{N_{Pr}}\frac{d}{d\eta}[g^{r-1}g'].$$

(1.66)

1.8 Population Growth and Other Problems in Biological Sciences

As the life sciences become more and more mathematized the occurrence of nonlinear equations is to be expected. Lotka's useful work [49a] and those of Rashevsky [24] have a number of problems formulated and discussed. Comprehensive discussion of the conflicting populations growth problem, not only from the mathematical view, but also from the origin and significance will be found in Volterra [23] and Rashevsky [24].

A. CONFLICTING POPULATION GROWTH

Let the two variables n and m be the respective measures of the number of individuals in the two conflicting populations. Let m measure the population of species A, which preys upon the second species B, whose population is measured by n. If n is large then A with so much food will flourish so that m increases. As a consequence n will decrease so a period of starvation sets in for species A. Then as m decreases the prey begins to increase in numbers and a cycling occurs in population size. The governing equations are

$$\frac{dn}{dt} = an - bnm$$

$$\frac{dm}{dt} = -cm + enm$$

(1.67)

where a, b, c, e are population parameters. One should note the similar format of Eqs. (1.67) to those of chemical kinetics (see Section 1.4). Indeed both systems are capable of formulation as "birth and death" processes by means of probability theory.

If we set $x = en/c$, $y = bm/a$ Eq. (1.67) becomes

$$\frac{dx}{dt} = ax(1 - y), \qquad \frac{dy}{dt} = -cy(1 - x).$$

To eliminate y, and hence obtain an equation for x, differentiate both equations, eliminate y and y' to get

$$x\frac{d^2x}{dt^2} = \left(\frac{dx}{dt}\right)^2 - cx(1 - x)\frac{dx}{dt} + acx^2(1 - x)$$

(1.68)

an equation of some difficulty.

Equations (1.67) can be and are modified to include such effects as heredity, other environmental factors, wars, disease, and so forth.

B. DIFFUSION AND REACTION IN BIOLOGICAL SYSTEMS

Trickling filters and activated sludge processes are extensively used for sewage treatment. Both poorly understood processes are biochemical in nature. The research of various sanitary engineers indicates that the processes, especially the biochemical reactions, are nonlinear. See for example Behn *et al.* [50] and Ames *et al.* [51]. The steady operation of the trickling filter has been characterized by Ames *et al.* [52] by using a diffusion, "logistic" reaction. The equation is

$$D\frac{d^2C}{dr^2} + \frac{2D}{r}\frac{dC}{dr} - kC(C_0 - C) = 0$$

(1.69)

where C, D, k, C_0, and r are respectively concentration, diffusion constant, reaction rate constant, limiting concentration, and distance.

C. BACTERIAL GROWTH

Investigations into the theory of the continuous bacterial growth apparatus have been carried out by Moser [53]. His basic equations are

$$
\frac{dN}{dt} = N[k(c) - v(c) - \omega]
$$
$$
\frac{dc}{dt} = \omega(a - c) - gk(c)N
$$

(1.70)

where c is the concentration of limiting growth factor in the culture vessel, k is the growth rate factor, v is the inviability factor, ω is the washout rate, a is the concentration of c in the incoming medium, g is the uptake of c per bacterium produced, and N is the population density. Moser asserts that k and v are indeed nonlinear but he does not investigate them further, except for the linear case $k(c) = \alpha c$ and $v(c) = 0$. The system, Eqs. (1.70), is still nonlinear, but solvable exactly for this case.

1.9 Radiation Heat Transfer

As technology expands and man reaches for the stars the temperature extremes encountered require a more thorough understanding of radiative heat transfer (see Hottel and Sarofim [54] for example). Since radiation is a fourth power phenomenon the basic equations are nonlinear. Typical of these is the idealized case of transient heat flow in an inert gas filled electron tube shown in Fig. 1-1. Initially, at $t < 0$, the whole system is at temperature T_a. At $t = 0$ the filament temperature is suddenly raised to $T_f > T_a$ by an electric current. Heat is convected to the surrounding gas and also radiated to the tube wall. The wall receives heat by convection from the gas and by radiation from the filament. Finally we assume the wall transfers heat by convection to the surrounding atmosphere at temperature T_a. The dimensionless equations for this problem (see Crandall

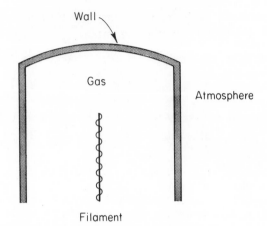

Fig. 1-1. Schematic of a radiation problem.

[55]) are

$$\frac{dx_1}{dt} = -2x_1 + x_2 + 2$$

$$\frac{dx_2}{dt} = \frac{1}{2}x_1 - x_2 + \frac{1}{2} + \frac{16 - x_2^4}{10}$$

(1.71)

$x_1(0) = 1, \ x_2(0) = 1.$

We note that some relations between physical parameters have been assumed to achieve Eqs. (1.71).

REFERENCES

1. Lane, J. H., *Am. J. Sci.* **50**, 57 (1870).
2. Emden, V. R., "Gaskugeln," Springer, Berlin, 1907.
3. Kelvin, Lord, *Proc. Roy. Soc. Edinburgh* **27A**, 375 (1907).
4. Chandrasekhar, S., "Introduction to the Study of Stellar Structure," Univ. Chicago Press, Illinois, 1939 (reprint: Dover, New York, 1961).
5. Davis, H. T., "Introduction to Nonlinear Differential and Integral Equations," Dover, New York, 1962.
6. Andronow, A., and Chaikin, S., "Theory of Oscillations." Moscow, 1937 (transl.: Princeton Univ. Press, Princeton, New Jersey, 1949).
7. Stoker, J. J., "Nonlinear Vibrations in Mechanical and Electrical Systems," Wiley (Interscience), New York, 1950.

8. McLachlan, N. W., "Ordinary Non-Linear Differential Equations in Engineering and Physical Sciences," 2nd ed., Oxford Univ. Press, London and New York, 1955.

9. Kryloff, N. M., and Bogoliuboff, N. N., "Introduction to Nonlinear Mechanics," Princeton Univ. Press, Princeton, New Jersey, 1943.

10. Minorsky, N., "Introduction to Non-Linear Mechanics." Edwards, Ann Arbor, Michigan, 1947.

11. Minorsky, N., and Leimanis, E., "Dynamics and Nonlinear Mechanics." Wiley, New York, 1958.

12. Minorsky, N., "Nonlinear Oscillation," Van Nostrand, Princeton, New Jersey, 1962.

13. Cunningham, W. J., "Introduction to Nonlinear Analysis." McGraw-Hill, New York, 1958.

14. Malkin, I. G., "Some Problems in the Theory of Nonlinear Oscillations," books 1, 2 (translated from publ. of State Publishing House of Tech. Theoret. Lit. Moscow, U. S. At. Energy Comm., Washington, D.C., 1956).

15. Hale, J. K., "Nonlinear Oscillations," McGraw-Hill, New York, 1963.

16. Lasalle, J. P., and Lefschetz, S., "Stability by Liapunov's Direct Method with Applications," Academic Press, New York, 1961.

17. Lasalle, J. P., and Lefschetz, S., eds., "Nonlinear Differential Equations and Nonlinear Mechanics," Academic Press, New York, 1963.

18. Lefschetz, S., "Differential Equations: Geometric Theory," Wiley (Interscience), New York, 1957.

19. Bickley, W. G., *Phil. Mag.* **17**, 603 (1934).

20. von Karman, Th., The engineer grapples with nonlinear problems, *Bull. Am. Math. Soc.* **46**, 615 (1940) (contains 178 refs.).

21. Ames, W. F., "Nonlinear Partial Differential Equations in Engineering," Academic Press, New York, 1965.

22. Schlichting, H., "Boundary Layer Theory," McGraw-Hill, New York, 1955.

23. Volterra, V., "Lecons sur la Theorie Mathematiques de la Lutte pour la Vie," Paris, 1931 [transl.: Dover, New York, 1957].

24. Rashevsky, N., "Mathematical Biophysics—Physico Mathematical Foundations of Biology," 3rd ed., Vol. I, II, Dover, New York, 1960.

25. Struble, R. A., "Nonlinear Differential Equations," McGraw-Hill, New York, 1962.

26. Saaty, T. L., and Bram, J., "Nonlinear Mathematics," McGraw-Hill, New York, 1964.

27. Benson, S., "Foundations of Chemical Kinetics," McGraw-Hill, New York, 1960.

28. Ames, W. F., *Ind. Eng. Chem.* **52**, 517 (1960).

29. Ames, W. F., *Ind. Eng. Chem. Fundamentals* **1**, 214 (1962).

30. Richardson, O. W., "The Emission of Electricity from Hot Bodies," Longmans Green, New York, 1921.

31. Walker, G. W., *Proc. Roy. Soc.* **A91**, 410 (1915).

32. Bateman, H., "Partial Differential Equations of Mathematical Physics," Cambridge Univ. Press, London and New York, 1959.
33. Chambré, P. L., *J. Chem. Phys.* **20**, 1795 (1952).
34. Ames, W. F., ed., Ad hoc exact techniques for nonlinear partial differential equations, *In* "Nonlinear Partial Differential Equations—Methods of Solution", Academic Press, New York, 1967.
35. Burgers, J. M., *Advan. Appl. Mech.* **1**, 171 (1948).
36. Hopf, E., *Commun. Pure Appl. Math.* **3**, 201 (1950).
37. Cole, J. D., *Quart. Appl. Math.* **9**, 225 (1951).
38. Oplinger, D. W., *J. Acoust. Soc. Am.* **32**, 1529 (1960).
39. Oplinger, D. W., *Proc. Intern. Congr. Rheol.*, *4th*, *Providence, 1963*, Pt.2, p. 231, Wiley (Interscience), New York, 1965.
40. Smith, P., *Appl, Sci. Res.* **A12**, 66 (1963).
41. Tomotika, S., and Tamada, K., *Quart. Appl. Math.* **7**, 381 (1949).
42. Keller, J. B., *Quart. Appl. Math.* **14**, 171 (1956).
43. Birkhoff, G., "Hydrodynamics," Princeton Univ. Press, Princeton, New Jersey, 1960.
44. Hansen, A. G., "Similarity Analyses of Boundary Value Problems in Engineering," Prentice–Hall, Englewood Cliffs, New Jersey, 1964.
45. Lee, S. Y., and Ames, W. F., Similarity solutions for non–Newtonian fluids, *A.I.Ch.E. J.* **12**, 700 (1966).
46. Acrivos, A., Shah, M. J., and Peterson, E. E., *A.I.Ch.E. J.* **6**, 312 (1960).
47. Gutfinger, C., and Shinnar, R., *A.I.Ch.E. J.* **10**, 631 (1964).
48. Kapur, J. N., *J. Phys. Soc. Japan* **17**, 1303 (1962).
49. Kapur, J. N., and Srivastava, R. C., *Z. Angew. Math. Phys.* **14**, 383 (1963).
49a. Lotka, A. J. "Elements of Physical Biology," Williams and Wilkins Co., Baltimore, Maryland, 1925.
50. Behn, V. C., Ames, W. F., and Keshavan, K., *J. Sanit. Eng. Div. Am. Soc. Civil Engrs.* **88**, SA1, 31 (1962).
51. Ames, W. F., Behn, V. C., and Collings, W. Z., *J. Sanit Eng. Div. Am. Soc. Civil Engrs.* **88**, SA3, 21 (1962).
52. Ames, W. F., Collings, W. Z., and Behn, V. C., Diffusion and reaction in the biota of the trickling filter, Tech. Rep. No. 19, Dept. Mech. Eng., Univ. of Delaware, Newark, Delaware, October 1962.
53. Moser, H., *Proc. Nat. Acad. Sci. U.S.* **43**, 222 (1957).
54. Hottel, H. C., and Sarofim, A. F., "Radiative Transport," McGraw-Hill, New York, 1965.
55. Crandall, S. H., "Engineering Analysis," McGraw–Hill, New York, 1956.

2

EXACT METHODS OF SOLUTION

Introduction

By exact methods we shall mean any procedure which enables the solution to be obtained by one or more quadratures of a function of one variable only. Thus we include any methods leading to $y = \int f(x)\,dx$ or $x = \int g(y)\,dy$ where x and y are independent and dependent variables, respectively. Not included are quadratures of $dy/dx = f(x, y)$.

The inapplicability of many of the methods of linear analysis leaves the analyst with few tools. Nevertheless, a number of procedures have been devised for obtaining exact solutions to special cases. Many of these equations are tabulated in the collections of Kamke [1] and Murphy [2]. Some of these methods can be motivated and suggestions for their generalization given. Many of the developments have a substantial "art" content. Imagination, ingenuity, and good fortune play an important role.

An exact analytic solution in algebraic form (even one that is approximate) allows information to be obtained relatively easily concerning the effects that parameter changes have. Thus it is usually worthwhile to expend considerable effort attempting to find an exact solution. This is especially true if more than a single solution is desired. For a single application of a numerical method gives only a *single* solution for one set of numerical parameters and auxiliary conditions. A change in any of these quantities requires the performance of an entirely new computation.

FIRST ORDER EQUATIONS

In this subdivision of the chapter we shall consider methods for the solution of the differential equation

$$F(x, y, y') = 0, \qquad y' = dy/dx \tag{2.1}$$

where F is a function that is continuous and usually, unless specified otherwise, possesses continuous first derivatives with respect to x, y, and y' in some domain of interest.

2.1 The Integrating Factor

We assume that Eq. (2.1) has the form

$$\frac{dy}{dx} = f(x, y) \tag{2.2}$$

with solution

$$u(x, y) = 0, \tag{2.3}$$

so that upon forming the differential du we have

$$\frac{\partial u}{\partial x} dx + \frac{\partial u}{\partial y} dy = 0. \tag{2.4}$$

From Eqs. (2.2) and (2.4) we can therefore write that

$$f(x, y) = -\frac{\partial u/\partial x}{\partial u/\partial y}. \tag{2.5}$$

Let us now suppose that $f(x, y)$ can be expressed as the quotient of two functions

$$f(x, y) = -p(x, y)/q(x, y) \tag{2.6}$$

so that Eq. (2.2) has the form

$$p(x, y) \, dx + q(x, y) \, dy = 0. \tag{2.7}$$

Comparing Eq. (2.7) with Eq. (2.4), and noting the relation $U_{xy} = U_{yx}$, we immediately observe that Eq. (2.7) is integrable provided

$$\frac{\partial p}{\partial y} = \frac{\partial q}{\partial x}. \tag{2.8}$$

If this be true we say that Eq. (2.7) is *exact* and the solution is obtained by partial integration of either

$$\frac{\partial u}{\partial x} = p(x, y),$$

or $\qquad\qquad\qquad\qquad\qquad\qquad\qquad\qquad\qquad$ (2.9)

$$\frac{\partial u}{\partial y} = q(x, y).$$

Using ∂x, ∂y to indicate partial integration we find that

$$u(x, y) = \int p(x, y) \, \partial x + r(y) \tag{2.10a}$$

or

$$u(x, y) = \int q(x, y) \, \partial y + s(x) \tag{2.10b}$$

where $r(y)$ and $s(x)$ are arbitrary functions of integration.

To determine $r(y)$ and $s(x)$ we substitute Eq. (2.10a) into $\partial u/\partial y = q(x,y)$ and obtain

$$\frac{dr}{dy} = q(x, y) - \frac{\partial}{\partial y} \int p(x, y) \, \partial x. \tag{2.11}$$

The left hand side of Eq. (2.11) is a function of y alone, so therefore the right side must also be a function of y alone. This is easily verified by noting that

$$\frac{\partial}{\partial x} \left\{ q(x, y) - \frac{\partial}{\partial y} \int p(x, y) \, \partial x \right\} = 0,$$

that is, a mere restatement of Eq. (2.8). We may therefore determine $r(y)$ by integration of Eq. (2.11) so that

$$r(y) = \int \left[q(x, y) - \frac{\partial}{\partial y} \int p(x, y) \, \partial x \right] dy. \tag{2.12}$$

In a similar fashion we find

$$s(x) = \int \left[p(x, y) - \frac{\partial}{\partial x} \int q(x, y) \, \partial y \right] dx. \qquad (2.13)$$

The solutions obtained by this procedure are usually *implicit*. The following example is typical. Consider the equation

$$(x^2 - 4xy + 3y^2) \, dx + (6xy - 2x^2 - y^2) \, dy = 0.$$

With $p = x^2 - 4xy + 3y^2$, $q = 6xy - 2x^2 - y^2$ we have $\partial p/\partial y = -4x + 6y = \partial q/\partial x$, so that the equation is exact. Then

$$u(x, y) = \int (x^2 - 4xy + 3y^2) \, \partial x + r(y)$$

$$= x^3/3 - 2x^2y + 3xy^2 + r(y).$$

From Eq. (2.12) we find

$$r(y) = \int [6xy - 2x^2 - y^2 + 2x^2 - 6xy] \, dy = -y^3/3 + C.$$

The solution of the equation is thus

$$x^3/3 - 2x^2y + 3xy^2 - y^3/3 + C = 0.$$

Usually, however, Eq. (2.7) is not exact. In such instances it is theoretically always possible and sometimes practical to make the equation exact by introducing an *integrating factor* $\lambda(x, y)$ so that

$$\frac{\partial(\lambda p)}{\partial y} = \frac{\partial(\lambda q)}{\partial x}. \qquad (2.14)$$

Expanding Eq. (2.14) we see that $\lambda(x, y)$ must satisfy the partial differential equation

$$q \frac{\partial \lambda}{\partial x} - p \frac{\partial \lambda}{\partial y} + \lambda \left[\frac{\partial q}{\partial x} - \frac{\partial p}{\partial y} \right] = 0. \qquad (2.15)$$

If Eq. (2.7) possesses a solution then it is easy to see that there exists an infinite number of integrating factors (see for example Morris and Brown [3]). In fact if λ is an integrating factor and $u(x, y) = 0$ is any solution of Eq. (2.7) then $\lambda F(u)$ is an integrating factor of Eq. (2.7) for arbitrary differentiable F.

Determination of integrating factors is often difficult. Special cases do exist in which the factor can be found. A substantial number of these are given in Table 2.1. See also Section 2.20 and Table 2.5.

TABLE 2.1

INTEGRATING FACTORS

Equation	Integrating factor
1. $f(y/x)\,dx - dy = 0$	$\lambda = [y - xf(y/x)]^{-1}$
2. $p(x, y)\,dx + q(x, y)\,dy = 0$	$\lambda = [xp + yq]^{-1}$
$\quad$ p and q homogeneous of	
$\quad$ the same degree, i.e.,	
$\quad$ $p(\varepsilon x, \varepsilon y) = \varepsilon^n p(x, y)$	
$\quad$ $q(\varepsilon x, \varepsilon y) = \varepsilon^n q(x, y)$	
3. $[y + xg]\,dx - [x - yg]\,dy = 0$	$\lambda = [x^2 + y^2]^{-1}$
$\quad$ $g = g(x^2 + y^2)$	
4. If $q^{-1}[p_y - q_x] = g(x)$	$\lambda = \exp\left[\displaystyle\int^{} g\,dx\right]$
5. If $(p_y - q_x)/(q\Phi_x - p\Phi_y) = g(\Phi)$	$\lambda = \exp\left[\displaystyle\int^{} g(\Phi)\,d\Phi\right]$
$\quad$ $\Phi = \Phi(x, y)$ with first partial	
$\quad$ derivatives	
5a. $\Phi = xy$	
$\quad$ If $(p_y - q_x)/(qy - px) = g(xy)$	$\lambda = \exp\left[\displaystyle\int^{} g(xy)\,d(xy)\right]$
5b. $\Phi = y/x$	
$\quad$ If $-x^2(p_y - q_x)/(yq + px) = g(y/x)$	$\lambda = \exp\left[\displaystyle\int^{} g(y/x)\,d(y/x)\right]$
6. If $p = yf(xy)$, $q = xg(xy)$	$\lambda = [xp - yq]^{-1}$
7. If $p_x = q_y$, $p_y = -q_x$	$\lambda = (p^2 + q^2)^{-1}$
8. $[f(x) + yg(x)]\,dx + y\,dy = 0$	$\lambda = \exp\left[-k\left(y + \displaystyle\int^{} g\,dx\right)\right]$
$\quad$ $g(x) = kf(x)$	
9. $[yf(x) - g(x)]\,dx + dy = 0$	$\lambda = \exp\left[\displaystyle\int^{} f(x)\,dx\right]$
10. $y[a + \alpha x^m y^n]\,dx + x[b + \beta x^m y^n]\,dy = 0$	$\lambda = x^\rho y^\eta$
$\quad$ $b\alpha - \beta a \neq 0$	ρ and η from
	$b\rho - a\eta = a - b$
	$\beta\rho - \alpha\eta = \alpha(n + 1) - \beta(m + 1)$
11. $p_y - q_x = pf(y) - qg(x)$	$\lambda = u(x)v(y)$
	$u' + gu = 0$
	$v' + fv = 0$
12. $(p_y - q_x - nqx^{-1})/(qu_x - pu_y) = \Phi(u)$	$\lambda = x^n f(u)$

2.2 Homogeneous Equations

If $f(x, y)$ satisfies the relation

$$f(\varepsilon x, \varepsilon y) = \varepsilon^n f(x, y) \qquad (2.16)$$

we say that it is homogeneous of degree n. A theorem due to Euler states that if $f(x, y)$ is homogeneous of degree n, then it satisfies the partial differential equation

$$xf_x + yf_y = nf \qquad (2.17)$$

where the subscripts indicate partial differentiation. An extension of this theorem is also easily proved. If $f(x, y: u, v)$ is homogeneous of degree m in x and y and of degree n in u and v then f satisfies the equation

$$(n - m)f = \left(u\frac{\partial}{\partial x} + v\frac{\partial}{\partial y}\right)\left(x\frac{\partial f}{\partial u} + y\frac{\partial f}{\partial v}\right) - \left(x\frac{\partial}{\partial u} + y\frac{\partial}{\partial v}\right)\left(u\frac{\partial f}{\partial x} + v\frac{\partial f}{\partial y}\right).$$
$$(2.18)$$

We now suppose that $p(x, y)$ and $q(x, y)$ are homogeneous functions of the same degree n. From Table 2.1, number 2 it is apparent that

$$\lambda(x, y) = [xp + yq]^{-1}$$

is an integrating factor for

$$p(x, y)\, dx + q(x, y)\, dy = 0. \qquad (2.19)$$

The solution of Eq. (2.19) is now obtained by setting

$$y = vx \qquad (2.20)$$

so that $dy = x\, dv + v\, dx$. Upon substituting these and utilizing the homogeniety of p and q we find

$$\begin{aligned}
0 &= p(x, y)\, dx + q(x, y)\, dy \\
&= p(x, vx)\, dx + q(x, vx)\,(x\, dv + v\, dx) \\
&= x^n p(1, v)\, dx + x^n q(1, v)\,(x\, dv + v\, dx).
\end{aligned}$$

After regrouping and division by x^n $(x \neq 0)$ we have

$$[p(1, v) + vq(1, v)]\, dx + xq(1, v)\, dv = 0 \qquad (2.21)$$

which may be separated as

$$\frac{dx}{x} + \frac{q(1, v)}{p(1, v) + vq(1, v)}\, dv = 0. \tag{2.22}$$

Its integral is

$$x = k \exp\left[-\int \frac{q(1, v)\, dv}{p(1, v) + vq(1, v)}\right] \tag{2.23}$$

where k is an integration constant. When the integration is completed replace v by y/x.

An important special equation of nonlinear mechanics can be transformed to a homogeneous equation. Stability analysis depends heavily upon the equation

$$(Ax + By + E)\, dx = (Cx + Dy + F)\, dy \tag{2.24}$$

where $AD - BC \neq 0$.† In its present form Eq. (2.24) is not homogenous but can reduced to homogeneous form by means of the linear transformation

$$x = x' + h, \qquad y = y' + k. \tag{2.25}$$

Setting these into Eq. (2.24) there results

$$(Ax' + By' + Ah + Bk + E)\, dx' = (Cx' + Dy' + Ch + Dk + F)\, dy'.$$

Values of h and k can be determined so that

$$Ah + Bk + E = 0$$

$$Ch + Dk + F = 0$$

provided $AD - BC \neq 0$. By this choice of h and k Eq. (2.24) is reduced to the homogeneous equation

$$(Ax + By)\, dx - (Cx + Dy)\, dy = 0 \tag{2.26}$$

where we have dropped the primes.

---

† If $AD - BC = 0$ then $Ax + By = Ax + (AD/C)\, y = (A/C)\, (Cx + Dy)$ and $Ax + By$ and $Cx + Dy$ are linearly dependent. By setting $u = Ax + By$ the equation is easily separated.

Equation (2.26) has the immediate solution

$$x = k \exp\left[\int^v \frac{C + Dv}{A + (B - C)v - Dv^2} \, dv\right].$$ (2.27)

Example

The evaluation of rate constants in reaction kinetics is a problem of considerable difficulty for complex reactions. Several cases have been treated by Ames [4, 5]. With his methods ratios of the rate constants and not the constants themselves are computable.

We consider the reaction

$$A_1 + A_2 \xrightarrow{k_1} A_3$$
$$A_3 + A_2 \xrightarrow{k_2} A_4$$
$$A_5 + A_2 \xrightarrow{k_3} A_6$$
$$A_6 + A_2 \xrightarrow{k_4} A_7$$

(2.28)

where reactant A_1 has an effective constant concentration C because it is in excess, goes into solution as the reaction progresses and therefore C moles per unit volume are always present.

The equations for the constant volume reaction are

$$dx_2/dt = -x_2[k_1C + k_2x_3 + k_3x_5 + k_4x_6]$$ (2.29a)

$$dx_3/dt = k_1Cx_2 - k_2x_2x_3$$ (2.29b)

$$dx_4/dt = k_2x_2x_3$$ (2.29c)

$$dx_5/dt = -k_3x_2x_5$$ (2.29d)

$$dx_6/dt = k_3x_2x_5 - k_4x_2x_6$$ (2.29e)

$$dx_7/dt = k_4x_2x_6$$ (2.29f)

where x_i is the concentration of A_i in moles per unit volume and k_i are the rate constants in liters per gram mole second. From material balances we find the additional relations

$$x_5 = x_5(0) - x_6 - x_7.$$ (2.30a)

$$x_2 = x_2(0) - x_3 - 2x_4 - x_6 - 2x_7.$$ (2.30b)

These redundancies are also obtainable from the equations. For example, upon adding Eqs. (2.29d, e, and f) and integrating we find Eq. (2.30a).

Equations (2.29) are nonlinear autonomous equations so the variable t may be eliminated by dividing one equation by another. If this is done judiciously solvable equations may result. Thus if one divides Eq. (2.29b) by Eq. (2.29c) there results

$$dx_3/dx_4 = (a/x_3) - 1 \qquad (2.31)$$

where $a = k_1 C/k_2$. Since Eq. (2.31) is separable its solution is immediately obtained as

$$a\{1 - \exp[-(x_3 + x_4)/a]\} - x_3 = 0 \qquad (2.32)$$

which is an implicit equation for the rate constant ratio as a function of x_3 and x_4.

A second ratio is obtainable by noting that Eq. (2.29d) integrates to

$$k_3^{-1} \ln[x_5(0)/x_5] = \int_0^t x_2 \, dt$$

and the sum of Eqs. (2.29b) and (2.29c) integrates to

$$(k_1 C)^{-1} (x_3 + x_4) = \int_0^t x_2 \, dt.$$

Upon equating these two relations we get an explicit relation

$$\frac{k_3}{k_1} = \frac{C}{x_3 + x_4} \ln\left[\frac{x_5(0)}{x_5}\right]. \qquad (2.33)$$

To obtain the third and final ratio k_4/k_3 we note that Eqs. (2.29d) and (2.29f) can be written as

$$x_6^{-1} \, dx_7/dt = k_4 x_2 = -(k_4/k_3) \, x_5^{-1} \, dx_5/dt. \qquad (2.34)$$

From Eq. (2.30a), $x_6 = x_5(0) - x_5 - x_7$, so that Eq. (2.34) becomes

$$x_5 \, dx_7 + b[x_5(0) - x_5 - x_7] \, dx_5 = 0, \qquad (2.35)$$

an equation of the form Eq. (2.24). Utilizing that theory the solution for $b = k_4/k_3$ is the implicit relation

$$b - 1 + x_6^{-1}\{x_5(0) [x_5/x_5(0)]^b - x_5\} = 0. \qquad (2.36)$$

2.3 General First Order Equations

In addition to those already considered some forms of the general nonlinear first order equation

$$F(x, y, y') = 0 \tag{2.37}$$

may be solved exactly. If Eq. (2.37) is not of the first degree in y' it may be satisfied by a multiplicity of values of y' for a pair of values (x, y).

A. EQUATIONS SOLVABLE FOR y'

If Eq. (2.37) is solvable for y' we may find several values in the form

$$y' = f_1(x, y), \quad y' = f_2(x, y), \quad \dots.$$

Further, each of these may be treated by preceding methods.

B. EQUATIONS SOLVABLE FOR y

Let Eq. (2.37) be solvable for y so that one or more expressions of the form

$$y = f(x, y') \tag{2.38}$$

are obtained. Differentiating this with respect to x we obtain

$$y' = f_x + f_{y'} \frac{dy'}{dx} \tag{2.39}$$

which is a differential equation of first order and first degree in x and y'. Let us write the solution as

$$g(x, y', c) = 0. \tag{2.40}$$

Having found this solution we may eliminate y' between Eqs. (2.38) and (2.40) to obtain the desired solution of Eq. (2.38).

As an example we consider $xy = (y')^2$. Then $y = x^{-1}(y')^2$, $y' = -x^{-2}(y')^2 + 2x^{-1}y'\, dy'/dx$ which factors to $y' = 0$ and $(dy'/dx) - (1/2x)y' = x$. The second integrates to $y' = \frac{2}{3}x^2 + cx^{1/2}$. From the first by eliminating y' we get the solution $y = 0$. From the second $xy = (\frac{2}{3}x^2 + cx^{1/2})^2$.

C. EQUATIONS SOLVABLE FOR x

In the event that Eq. (2.37) is solvable for x we get one or more expressions of the form

$$x = f(y, y'). \qquad (2.41)$$

This equation is then differentiated with respect to y so that

$$\frac{dx}{dy} = f_y + f_{y'} \frac{dy'}{dy}$$

or

$$\frac{1}{y'} = f_y + f_{y'} \frac{dy'}{dy} \qquad (2.42)$$

which is an equation in y' and y. Having solved this equation we eliminate y' between Eqs. (2.41) and the solution.

As an example we consider $x = 2y' - (y')^2$ so that $(y')^{-1} = 2(1 - y')\, dy'/dy$ which integrates to $(y')^2 - \frac{2}{3}(y')^3 = y$. The solution is obtained by eliminating y' between the two equations.

D. OTHER EQUATIONS SOLVABLE BY DIFFERENTIATION

In addition to those equations discussed in Sections B and C, which were solvable by differentiation, we add Clairaut's equation

$$y = xy' + \varphi(y'). \qquad (2.43)$$

Clairaut's equation falls into the category of an equation solvable for y. Upon differentiation with respect to x we get

$$y' = y' + [x + \varphi'(y')] \frac{dy'}{dx} \qquad (2.44)$$

which can only be satisfied if $dy'/dx = 0$ or

$$x = -\varphi'(y'). \qquad (2.45)$$

From $dy'/dx = 0$ we find $y' = C$ which combines with Eq. (2.43) to yield the solution

$$y = Cx + \varphi(C). \qquad (2.46)$$

Next, fixing our attention on Eq. (2.45), we see that by substituting it into Eq. (2.43) there results

$$y = -y'\varphi'(y') + \varphi(y'). \tag{2.47}$$

Evidently Eqs. (2.45) and (2.47) may be regarded as the parametric equations of a curve that is an integral curve (solution) of the equation. Such solutions bear the name *singular solutions*. They are further discussed by Kamke [1] and Ince [6] for example.

2.4 Solution by Transformation

A large number of nonlinear ordinary differential equations while not exactly solvable in their original form may be solvable after transformation. A substantial number of these are tabulated by Kamke [1] and Murphy [2]. Here we discuss the basic ideas remarking first that such transformations may be on the dependent variable alone or on a combination of both dependent and independent variables.

A. On the Dependent Variable

(i)   Let $f(x)$, $g(x)$ be continuous functions of x and $n \neq 0$ or 1. The *Bernoulli equation*

$$\frac{dy}{dx} + f(x)y = g(x)y^n \tag{2.48}$$

has arisen in several engineering problems. Among these we find those of a capacitor C discharging through a nonlinear diode where the instantaneous values of current i and voltage e are expressed as

$$i = ae + be^n, \tag{2.49}$$

with a and b positive constants. The resulting equation for the circuit is

$$C\frac{de}{dt} + ae + be^n = 0. \tag{2.50}$$

A special case of this example as well as application to a delay oscillator and a negative resistance oscillator is discussed by Cunningham [7].

We attempt to transform Eq. (2.48) to an integrable form in a new dependent variable u where

$$y = F(u). \tag{2.51}$$

Upon setting $dy/dx = F'(u) \, du/dx$ and $y = F(u)$ into Eq. (2.48) we find

$$F'(u) \frac{du}{dx} + f(x)F(u) = g(x)F''(u)$$

or

$$\frac{du}{dx} + f(x) \frac{F(u)}{F'(u)} = g(x) \frac{F''(u)}{F'(u)}. \tag{2.52}$$

Our twin goals are to find $F(u)$ so that Eq. (2.52) is of the general linear form and hence integrable. Thus we wish

$$F/F' = \alpha, \qquad F''/F' = \beta \tag{2.53}$$

for some constants α and β. The solutions of these equations

$$F = Au^{1/\alpha}, \qquad F = Bu^{1/(1-n)} \tag{2.54}$$

are compatible if $\alpha = 1 - n$ and $A = B$. Without loss of generality, we take $A = B = 1$ and the transformation becomes

$$y = F(u) = u^{1/(1-n)}. \tag{2.55}$$

With this $F(u)$ Eq. (2.52) takes the form

$$\frac{du}{dx} + (1 - n)f(x) u = (1 - n) g(x)$$

whose solution is

$$u = \exp\left[(n - 1) \int f(x) \, dx \right] \left\{ A - (n - 1) \int g(x) \exp\left[(1 - n) \int f(x) \, dx \right] dx \right\}. \tag{2.56}$$

(ii) The nonlinear differential equation

$$\frac{dy}{dx} + q(x) y + r(x) y^2 = p(x), \tag{2.57}$$

called the *Riccati equation* occurs in an intermediate way in various problems of engineering. We assume $p(x)$ is not identically zero, for in this case we obtain a Bernoulli equation with $n = 2$.

A simple example of the application of a form of the Riccati equation occurs in the problem of a body of mass m falling in a gravity field in a medium which offers resistance to the motion proportional to the square of the velocity. The equation of motion is

$$m \frac{d^2s}{dt^2} = mg - k\left(\frac{ds}{dt}\right)^2 \tag{2.58}$$

or

$$m \frac{dv}{dt} = mg - kv^2 \tag{2.59}$$

where s, v, and t are distance, velocity, and time, respectively. Equation (2.59) is a Riccati equation. The physical problem, together with experiments, associated with Eqs. (2.58) and (2.59) is discussed by Davis [8].

Yet a second reason for the importance of the Riccati equation is its relationship to the general homogeneous linear differential equation of second order

$$\frac{d^2u}{dx^2} + b(x)\frac{du}{dx} + C(x)\,u = 0. \tag{2.60}$$

If we introduce an arbitrary function $R(x)$ and make the transformation

$$\frac{du}{dx} = \{R(x)y\}u \tag{2.61}$$

we obtain the Riccati equation

$$\frac{dy}{dx} + \left(\frac{R'}{R} + b\right)y + Ry^2 = -\frac{C}{R}. \tag{2.62}$$

Upon comparison of Eq. (2.62) with Eq. (2.57) we can write

$$r(x) = R(x), \qquad q(x) = \frac{R'(x)}{R(x)} + b(x), \qquad p(x) = -\frac{C(x)}{R(x)}. \tag{2.63}$$

But, since $R(x)$ was arbitrary we can determine it so as to force $q(x)$ to be zero. Thus with

$$R(x) = \exp\left[-\int b(x)\,dx\right]$$

we find that y satisfies the simpler form

$$\frac{dy}{dx} + Ry^2 = -\frac{C}{R}. \tag{2.64}$$

A detailed analysis for Eq. (2.57) is given by Ince [6] and Davis [8]. We omit the general form herein and discuss only the integrable form first given by Riccati in 1724,

$$\frac{dy}{dx} + ay^2 = bx^n \tag{2.65}$$

where a, b, and n are real constants. From Eq. (2.61) we have

$$u = \exp\left[a \int y \, dx\right], \qquad u' = ayu, \qquad u'' = [ay' + a^2y^2]u$$

whereupon Eq. (2.65) transforms to

$$\frac{d^2u}{dx^2} - abx^nu = 0 \tag{2.66}$$

which is a Bessel equation (see Kamke [1]). The solution of

$$\frac{d^2u}{dx^2} - \alpha x^{-1}\frac{du}{dx} + \beta^2 x^\mu u = 0 \tag{2.67}$$

is

$$u = x^p[AJ_\nu(kx^q) + BJ_{-\nu}(kx^q)] \tag{2.68}$$

where

$p = (\alpha + 1)/2$, $\nu = (\alpha + 1)/(\mu + 2)$, $k = 2\beta/(\mu + 2)$, $q = p/\nu = (\mu + 2)/2$,

and ν is not an integer. Eq. (2.66) takes the form of Eq. (2.67) if $\alpha = 0$ and $\beta^2 = -ab$. Thus $p = \frac{1}{2}$, $\nu = 1/(\mu + 2)$, $k = 2i(ab)^{1/2}/(\mu + 2)$, $q = (\mu + 2)/2$. Since k is imaginary the modified Bessel function

$$I_\nu(Z) = \exp(-\nu\pi i/2)J_\nu(iZ)$$

can be introduced into Eq. (2.68). Thus the solution of Eq. (2.66) is expressed as

$$u = x^{1/2}[A_1I_\nu(\gamma x^q) + B_1I_{-\nu}(\gamma x^q)] \tag{2.69}$$

where $\gamma = 2(ab)^{1/2}/(\mu + 2)$ and A_1, B_1 are integration constants. Since $y = u'/au$ we have

$$y = \frac{1}{2ax}\left\{1 + \frac{2\gamma qx^q[I_v' + CI_{-v}']}{I_v + CI_{-v}}\right\} \qquad (2.70)$$

where $C = B_1/A_1$ is the single integration constant.

(iii)   Actually the complete solution of a Riccati equation is generally attainable only by the integration of a linear differential equation of second order, or by some equivalent method. If, however, any particular solution is known, it is possible to develop the complete solution by means of guadratures. Let $y = Y$ be a particular integral of Eq. (2.57). Upon substituting

$$y = u^{-1} + Y \qquad (2.71)$$

into Eq. (2.57) we have

$$u^{-2}\left[-\frac{du}{dx} + qu + r + 2Yru\right] + \frac{dY}{dx} + qY + rY^2 = p.$$

Since Y is a solution of Eq. (2.57) the foregoing equation reduces to the first order linear equation

$$\frac{du}{dx} - [2Yr + q]u = r \qquad (2.72)$$

which is immediately integrable. Similarily, if two particular solutions are known the general solution can also be obtained.

B. On a Combination of Both Variables

(i)   A simple example of this idea is provided by the homogeneous equation

$$(x + y)\,dx = (x - y)\,dy. \qquad (2.73)$$

Upon introducing two new variables r, θ such that

$$x = r\cos\theta, \qquad y = r\sin\theta \qquad (2.74)$$

we find

$$dx = \cos\theta\,dr - r\sin\theta\,d\theta,$$
$$dy = \sin\theta\,dr + r\cos\theta\,d\theta. \qquad (2.75)$$

Upon setting Eq. (2.74) and Eq. (2.75) into Eq. (2.73) we find $dr/d\theta = r$ so that

$$\log r = \theta + A \tag{2.76}$$

with A the constant of integration. In the original variables Eq. (2.76) becomes

$$\tfrac{1}{2}\log(x^2 + y^2) - \tan^{-1}(y/x) = A \tag{2.77}$$

which is an implicit form.

(ii) A further application is that of a *transformation of Legendre*. Let us set $p = dy/dx$ and consider the general differential equation of first order

$$f(x, y, p) = 0. \tag{2.78}$$

Legendre introduced the new variable

$$Y = px - y \tag{2.79}$$

whose differential is

$$dY = p\,dx + x\,dp - dy = (p - y')\,dx + x\,dp = x\,dp. \tag{2.80}$$

If now, X is so chosen that $X = p$, then $dX = dp$. Thus from Eq. (2.80) we find

$$P = \frac{dY}{dX} = x. \tag{2.81}$$

The foregoing results transform the general first order differential equation, Eq. (2.78) into the *dual* equation

$$f(P, PX - Y, X) = 0 \tag{2.82}$$

which is likewise of first order. The solution of Eq. (2.82) is expressible as

$$F(X, Y, C) = 0. \tag{2.83}$$

To return to the original variables set $X = p$, $Y = px - y$ and obtain

$$F(p, px - y, C) = 0. \tag{2.84}$$

The solution to the original problem is then obtained by eliminating p between Eqs. (2.84) and (2.78). If this is not possible the equations for x and y may be left in parametric form, with parameter p.

For example the equation $xp^2 + 4p - 2y = 0$ transforms to

$$PX^2 + 4X - 2(PX - Y) = 0$$

under the Legendre transformation. This is clearly linear and therefore solvable by standard methods with the result

$$Y = (CX - 4X \ln X)/(X - 2).$$

Setting $X = p$, $Y = px - y$ this transforms to

$$y = px + (4p \ln p - Cp)/(p - 2).$$

From the original equation, $x = (2y - 4p)/p^2$. These may be manipulated to obtain parametric forms for x and y.

2.5 Further Solution by Differentiation

Consider the general first order equation

$$f(x, y, p) = 0, \qquad p = y'. \tag{2.85}$$

Let us regard x and y as functions of the parameter $t = p$. Differentiating with respect to t we find

$$\frac{\partial f}{\partial t} + \frac{\partial f}{\partial x}\frac{dx}{dt} + \frac{\partial f}{\partial y}\frac{dy}{dt} = 0$$

or

$$\frac{\partial f}{\partial t} + \left[\frac{\partial f}{\partial x} + \frac{\partial f}{\partial y}\frac{dy}{dx}\right]\frac{dx}{dt} = 0.$$

Using the convenient subscript notation for partial derivatives we rewrite the last equation as

$$\frac{dx}{dt} = -f_t/(f_x + tf_y). \tag{2.86}$$

Similarily we find

$$\frac{dy}{dt} = -tf_t/(f_x + tf_y) \tag{2.87}$$

where the results are invalid if $f_x + tf_y = 0$. Equations (2.86) and (2.87) constitute two equations for x and y. If these are easier to solve than Eq. (2.85) they result in parametric solutions for x and y.

A number of special cases are solvable by this method. We consider two not previously detailed.

(i) If $y = G(x, y')$ then Eq. (2.86) takes the form

$$\frac{dx}{dt} = \frac{G_t(x, t)}{t - G_x(x, t)}, \qquad t - G_x \neq 0. \tag{2.88}$$

If $x(t)$ is the solution of Eq. (2.88) then the parametric solution is

$$x = x(t), \qquad y = G(x(t), t). \tag{2.89}$$

(ii) If $x = G(y, y')$ then Eq. (2.87) becomes

$$\frac{dy}{dt} = \frac{tG_t(y, t)}{1 - tG_y(y, t)}, \qquad 1 - tG_y \neq 0. \tag{2.90}$$

The parametric solution is

$$x = G(y(t), t), \qquad y = y(t).$$

As an example consider the equation $yy'^2 - 4xy' + y = 0$. Here either form (i) or (ii) may be used. Using (ii), $x = (yy'^2 + y)/4y'$ so that

$$\frac{dy}{dt} = \frac{y(t^2 - 1)}{t(3 - t^2)}$$

whose solution is $y^3 = C/t(t^2 - 3)$. Thus the parametric solution is

$$y^3 = C/t(t^2 - 3), \qquad x = y(t^2 + 1)/4t.$$

SECOND ORDER EQUATIONS

Nonlinear second order equations occur more often in engineering than any other category. A close second, especially in fluid mechanics, is the third order equation. A variety of second order equations are solvable exactly. A number of these forms are given herein.

2.6 The Simplest Equations

A. ONLY SECOND DERIVATIVE PRESENT

Some of the simplest problems in engineering give rise to equations of the form

$$\frac{d^2 y}{dx^2} = f(y). \tag{2.91}$$

Typical engineering examples include the equation

$$\frac{d^2 \theta}{dt^2} + \left(\frac{g}{L}\right) \sin \theta = 0$$

for the oscillations of a pendulum (Davis [8]); oscillation of a nonlinear spring mass system is governed by

$$\frac{d^2 y}{dt^2} = ay + by^3,$$

an equation whose generalization is called the *Duffing equation* (Stoker [9]); the handle of fabrics utilizes the equation for the heavy elastica

$$\frac{d^2 \theta}{d\sigma^2} = -\sigma \cos \theta$$

treated by Bickley [10]; the problem of perihelion shift led Einstein to consider the equation

$$\frac{d^2 y}{dx^2} = a - y + by^2;$$

the equation for the thermal balance between the heat generated by a chemical reaction and that conducted away may be approximated by

$$\frac{d^2 \theta}{dZ^2} = -\delta \exp \theta, \tag{2.92}$$

an equation considered by numerous authors in this and generalized form (Chambré [11]). Equations of the form Eq. (2.92) also occur in the theory of the electric charge around a hot wire and in certain problems of solid mechanics.

To solve Eq. (2.91) a first integral is obtained by multiplying by the factor

$$2(dy/dx)\, dx = 2dy$$

and integrating to give

$$2\int \frac{d^2y}{dx^2}\frac{dy}{dx}\, dx = 2\int f(y)\, dy + C$$

or

$$\left(\frac{dy}{dx}\right)^2 = 2\int f(y)\, dy + C. \tag{2.93}$$

The square root is taken of each side of Eq. (2.93), the variables separated and a quadrature remains. A number of these integrations are possible in terms of elliptic functions as described in Section 2.7.

A first integral of Eq. (2.92) is easily obtained by the foregoing device. Multiplying Eq. (2.92) by $2(d\theta/dZ)\, dZ = 2d\theta$ and integrating we obtain $(d\theta/dZ)^2 = C - 2\delta \exp \theta.$

B. Missing Dependent Variable

If the differential equation does not contain the dependent variable y, explicitly, it would have the general form

$$f\left(\frac{d^2y}{dx^2}, \frac{dy}{dx}, x\right) = 0. \tag{2.94}$$

By setting $p = dy/dx$ the equation is reduced to the first order equation

$$f(dp/dx, p, x) = 0. \tag{2.95}$$

This equation may be solvable for p by one of the preceding methods. Having accomplished this integration another first order equation, in y, remains.

Examples of this form of equation includes

$$\frac{w}{g}\frac{d^2y}{dt^2} = w - \lambda\left(\frac{dy}{dt}\right)^2 \tag{2.96}$$

for a body falling from rest in a medium offering resistance proportional to the square of the velocity (Davis [8], Carlson *et al.* [12]). The pursuit of prey, either biological or in warfare has led to extensive studies of curves of pursuit. Hathaway [13] gives a detailed discussion of such problems. When the pursued moves along a straight line, the curve of pursuit calculation is accomplished by means of the equation

$$1 + \left(\frac{dy}{dx}\right)^2 = k^2(a - x)^2 \left[\frac{d^2y}{dx^2}\right]^2, \qquad k > 0. \qquad (2.97)$$

To solve Eq. (2.97), subject to the initial conditions $y = dy/dx = 0$ when $x = 0$, we set $p = dy/dx$ whereupon Eq. (2.97) becomes

$$\frac{k \, dp}{(1 + p^2)^{1/2}} = \frac{dx}{a - x}$$

from which we find by integration

$$\frac{dy}{dx} = \frac{1}{2}\left[C(a - x)^{-1/k} - \frac{1}{C}(a - x)^{1/k}\right]$$

where C is the integration constant. The positive root has been chosen for physical reasons. A second integration produces the result

$$y(x) = \frac{1}{2}\left[\frac{kC}{1 - k}(a - x)^{(k-1)/k} + \frac{k}{C(1 + k)}(a - x)^{(k+1)/k}\right] + C_1$$

if $k \neq 1$. From the initial conditions we find $C = a^{1/k}$, $C_1 = ka/(k^2 - 1)$ so that the solution is

$$y = \frac{ka}{k^2 - 1} + \frac{ka}{2(k^2 - 1)}$$

$$\times \left[(k - 1)\left(1 - \frac{x}{a}\right)^{(k+1)/k} - (k + 1)\left(1 - \frac{x}{a}\right)^{(k-1)/k}\right].$$

If $k = 1$ the solution is easily obtained as

$$y = \frac{a}{4}\left[\left(1 - \frac{x}{a}\right)^2 - \log\left(1 - \frac{x}{a}\right)^2 - 1\right].$$

C. MISSING INDEPENDENT VARIABLE

Occasionally the equation does not contain the independent variable x, explicitly. Hence the equation has the general form

$$f\left(\frac{d^2y}{dx^2}, \frac{dy}{dx}, y\right) = 0. \tag{2.98}$$

By setting $p = dy/dx$ we find that

$$\frac{d^2y}{dx^2} = \frac{dp}{dx} = \frac{dp}{dy}\frac{dy}{dx} = p\frac{dp}{dy} \tag{2.99}$$

whereupon Eq. (2.98) is transformed to a first order equation in p and y. This equation may be integrated by one of the first order methods. There then remains another first order equation in y with independent variable x.

A considerable number of engineering problems give rise to equations of this form. The development of similarity variables (see, e.g., Ames [14] or the Appendix) in fluid mechanics gives rise to a multiplicity of such equations although they are very often of third order (see Eqs. (1.62)–(1.66)). In addition Eq. (1.55), Eq. (1.68), and Eq. (1.69) are of this type.

Additional examples include the equation for the capillary curve

$$\frac{d^2y}{dx^2} = \frac{4y}{C^2}\left[1 + \left(\frac{dy}{dx}\right)^2\right]^{3/2}, \qquad C^2 = \frac{4T}{\rho g} \tag{2.100}$$

(Reddick and Miller [15]); the suspended cable (catenary) equation

$$\frac{d^2y}{dx^2} = \frac{w}{H}\left[1 + \left(\frac{dy}{dx}\right)^2\right]^{1/2};$$

the swinging cord equation

$$\frac{d^2y}{dx^2} + \frac{w\omega^2}{T}y\left[1 + \left(\frac{dy}{dx}\right)^2\right]^{1/2} = 0;$$

and the Van der Pol equation

$$\frac{d^2x}{dt^2} - \varepsilon(1 - x^2)\frac{dx}{dt} + x = 0.$$

Another equation of some interest is the Langmuir equation [16]

$$3y\frac{d^2y}{dx^2} + \left(\frac{dy}{dx}\right)^2 + 4y\frac{dy}{dx} + y^2 = 1$$

which appears in connection with the theory of current flow from a hot cathode to an anode in a high vacuum. The cathode and anode are long coaxial cylinders where r is the radius of the anode enclosing a cathode of radius r_0. The variables are $y = f(r/r_0)$ and $x = \log(r/r_0)$.

Additional examples are given by Ames [17].

In solving Eq. (2.100) we suppose the physical situation is as shown in Fig. 2-1. Thus $y(0) = l_0$ and $(dx/dy)(0) = -\cot \alpha$. We now set $p = y'$,

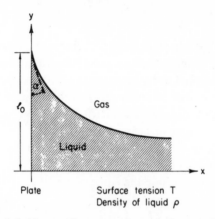

Fig. 2-1. The capillary curve for a single wetted wall.

$y'' = p \, dp/dy$ so that Eq. (2.100) becomes

$$\frac{p \, dp}{(1 + p^2)^{3/2}} = \frac{4y \, dy}{C^2}$$

which integrates to

$$-(1 + p^2)^{-1/2} = 2y^2/C^2 + C_1. \tag{2.101}$$

Since $p = y' = 0$ when $y = 0$ we have $C_1 = -1$ so that Eq. (2.101) becomes

$$(1 + p^2)^{-1/2} = \frac{C^2 - 2y^2}{C^2}. \tag{2.102}$$

At $x = 0$, $y = l_0$, and $p = -\cot \alpha$. Substitution of these values into Eq. (2.102) gives the expression

$$l_0 = C\left[\frac{1 - \sin \alpha}{2}\right]^{1/2} = C \sin\left[\frac{\pi}{4} - \frac{\alpha}{2}\right]$$

for l_0.

Now, returning to Eq. (2.102) we may square the reciprocal of both sides to obtain

$$1 + p^2 = \frac{C^4}{(C^2 - 2y^2)^2}$$

or

$$p = \frac{dy}{dx} = -\frac{2y(C^2 - y^2)^{1/2}}{C^2 - 2y^2}$$

where the negative sign is used since the physical slope of the curve in Fig. 2-1 is negative for a "wetting" fluid. Separating the variables we have

$$dx = \frac{y\,dy}{(C^2 - y^2)^{1/2}} - \frac{C^2}{2}\frac{dy}{y(C^2 - y^2)^{1/2}}$$

which integrates to

$$x + C_2 = -(C^2 - y^2)^{1/2} + (C/2)\,\text{sech}^{-1}(y/C).$$

But $y = l_0$ when $x = 0$ so that

$$C_2 = -(C^2 - l_0^2)^{1/2} + (C/2)\,\text{sech}^{-1}(l_0/C).$$

2.7 Elliptic Integrals

All the elementary functions can be considered to arise as solutions of certain differential equations. Thus $x = \exp bt$ is the solution of $dx/dt = bx$, $\sinh bt$ and $\cosh bt$ come from $d^2x/dt^2 - b^2x = 0$ and so forth. More complicated functions such as the Bessel functions, Legendre functions, etc also arise from obtaining series solutions to differential equations. Most of these functions have been extensively studies and are tabulated in a variety of references. A sampling of these would include Jahnke *et al.* [18], the National Bureau of Standards tables [19] and the "Bateman Manuscript Project" [20]. A list of pre-1962 tables is given in the detailed work of Fletcher *et al.* [21].

Many of the problems of the previous sections gave rise to equations requiring the integration of

$$\left(\frac{dy}{dx}\right)^2 = \sum_{n=0}^{N} a_n y^n. \qquad (2.103)$$

For $N \leq 2$ an integral can always be obtained in terms of the classical elementary functions (algebraic, trigonometric, inverse trigonometric, exponential, and logarithmic) even though this form may be very complicated. When N is 3 or 4 integration is performable in terms of functions called *elliptic functions*.

Originally the elliptic functions evolved out of attempts to find the arc-length of the ellipse $(x^2/a^2) + (y^2/b^2) = 1$ which may be expressed as $x = a \sin \Phi$, $y = b \cos \Phi$† in the parameter Φ. Arc length s from $\Phi = 0$ to $\Phi = \eta$ is then found by integrating

$$
\begin{aligned}
s = \int_0^s [dx^2 + dy^2]^{1/2} &= \int_0^\eta [a^2 \cos^2 \Phi + b^2 \sin^2 \Phi]^{1/2} \, d\Phi \\
&= a \int_0^\eta \left[1 - \frac{a^2 - b^2}{a^2} \sin^2 \Phi \right]^{1/2} \, d\Phi \\
&= a \int_0^\eta [1 - k^2 \sin^2 \Phi]^{1/2} \, d\Phi \qquad (2.104)
\end{aligned}
$$

where $k = (a^2 - b^2)^{1/2}/a$ is the eccentricity of the ellipse. It has been conclusively proven that Eq. (2.104) can not be evaluated in terms of elementary functions.

The integral in Eq. (2.104) is a function of two arguments, k called the *modulus* and η (upper limit) called the *amplitude*. We therefore write this *elliptic integral of the second kind* as

$$
E(k, \eta) = \int_0^\eta [1 - k^2 \sin^2 \Phi]^{1/2} \, d\Phi, \qquad k^2 < 1. \qquad (2.105)
$$

The total length of the ellipse is obtained from the *complete elliptic integral of the second kind* obtained by setting $\eta = \pi/2$,

$$
E(k, \pi/2) = \int_0^{\pi/2} [1 - k^2 \sin^2 \Phi]^{1/2} \, d\Phi.
$$

Thus the total length of the ellipse is $L = 4aE(k, \pi/2)$.

Other elliptic integrals are also fundamental. The *elliptic integral of*

---

† The major axis is the x axis, hence $a > b$. The parameter Φ has the range $0-\pi/2$.

the first kind, denoted by $F(k, \eta)$ is

$$F(k, \eta) = \int_0^{\eta} [1 - k^2 \sin^2 \Phi]^{-1/2} \, d\Phi, \qquad k^2 < 1. \qquad (2.106)$$

If in Eq. (2.106), known as *Legendre's form,* we set $u = \sin \Phi$, we get Jacobi's form of the elliptic integral of the first kind

$$F(k, \Phi) = F_1(k, x) = \int_0^x \frac{du}{[(1 - u^2)(1 - k^2 u^2)]^{1/2}}, \qquad k^2 < 1. \quad (2.107)$$

In a similar manner Legendre's form, Eq. (2.105), of the elliptic integral of the second kind transforms under $u = \sin \Phi$ to Jacobi's form

$$E(k, \Phi) = E_1(k, x) = \int_0^x \frac{(1 - k^2 u^2) \, du}{[(1 - u^2)(1 - k^2 u^2)]^{1/2}}, \qquad k^2 < 1. \quad (2.108)$$

Equations (2.107) and (2.108) establish a clear relationship between these elliptic integrals and integrals of the square root of polynomials, of at least order 4. In fact in each of these equations the integrand is a rational function of u and the square root of a quartic polynomial in u—that is an integral of the form

$$\int R(u, [a_0 u^4 + a_1 u^3 + a_2 u^2 + a_3 u + a_4]^{1/2}) \, du \qquad (2.109)$$

where R denotes a rational function of the two arguments. A few of these integrals can be evaluated in terms of elementary functions. In general Eq. (2.109) includes Eqs. (2.107) and (2.108) and *elliptic integrals of the third kind,* whose Jacobi form is

$$\pi_1(x, n, k) = \int_0^x \frac{du}{(1 + nu^2) [(1 - u^2)(1 - k^2 u^2)]^{1/2}}, \qquad k^2 < 1, \quad (2.110)$$

or its equivalent Legendre form $(u = \sin \Phi)$

$$\pi(\Phi, n, k) = \int_0^{\eta} \frac{d\Phi}{(1 + n \sin^2 \Phi) [1 - k^2 \sin^2 \Phi]^{1/2}}, \qquad k^2 < 1.$$

The graphs of the integrands of the Legendre forms of the first and second kinds of elliptic integrals are shown in Fig. 2-2. Both functions are periodic of period π. The function $y = [1 - k^2 \sin^2 \Phi]^{-1/2}$ oscillates between its minimum value unity and maximum value which depends upon k. Similarily $y = [1 - k^2 \sin^2 \Phi]^{1/2}$ has maximum value one and

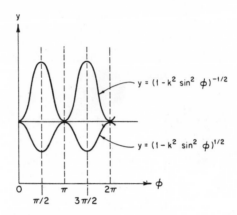

Fig. 2-2. Integrands of elliptic integrals of the first and second kinds.

TABLE 2.2

1. $(1 - k^2 y^2)\dfrac{d^2 y}{dx^2} - k^2 y \left(\dfrac{dy}{dx}\right)^2 + y = 0$ integrates to

$$\left(\dfrac{dy}{dx}\right)^2 = (1 - y^2)/(1 - k^2 y^2).$$

2. If $A > B > 0$ the integral $S = \displaystyle\int_0^x [(1 + Au^2)/(1 + Bu^2)]^{1/2}\, dx$

   has the value

   $$S = A^{-1/2}F[[(A - B)/A]^{1/2}, \Phi] - A^{1/2}B^{-1}\{E[[(A - B)/A]^{1/2}, \Phi] + \Delta\Phi \tan \Phi\}$$
   $$\tan \Phi = xA^{1/2}, \quad \Delta\Phi = [1 - (A - B)\sin^2 \Phi/A]^{1/2}.$$

3. $Z'' + k^2 \sin Z = 0$ transforms under $y = \sin(Z/2)$ to
   $$y'' = ay + by^3$$

4. $\displaystyle\int \frac{dZ}{[(1 - Z^2)(1 - k^2 Z^2)]^{1/2}} = \tfrac{1}{2}[(\beta - \delta)(\alpha - \gamma)]^{1/2}$

   $$\times \int \frac{dx}{[(x - \alpha)(x - \beta)(x - \gamma)(x - \delta)]^{1/2}}$$

   where

   $$Z^2 = \left(\frac{\beta - \delta}{\alpha - \delta}\right)\left(\frac{x - \alpha}{x - \beta}\right), \quad k^2 = \left(\frac{\beta - \gamma}{\alpha - \gamma}\right)\left(\frac{\alpha - \delta}{\beta - \delta}\right).$$

TABLE 2.2 (*Continued*)

5. $\int \dfrac{dZ}{[(1 - Z^2)(1 - k^2 Z^2)]^{1/2}} = -\tfrac{1}{2}(\alpha - \gamma)^{1/2} \int \dfrac{dy}{[(y - \alpha)(y - \beta)(y - \gamma)]^{1/2}}$

$$\text{where } Z^2 = \frac{\alpha - \gamma}{y - \gamma}, \qquad k^2 = \frac{\beta - \gamma}{\alpha - \gamma}.$$

6. $\int_0^\Phi \dfrac{du}{[1 - k^2 \sin^2 u]^{1/2}} = \dfrac{1}{k} F\!\left(\dfrac{1}{k}, x\right), \qquad k > 1$

where $x = \sin^{-1}(k \sin \Phi)$, integral is real valued for $k \sin \Phi \leq 1$.

7. $\int_0^\Phi \dfrac{\sin^2 u \, du}{[1 - k^2 \sin^2 u]^{1/2}} = \begin{cases} [F(k, \Phi) - E(k, \Phi)]/k^2, & k^2 < 1 \\[2mm] \dfrac{1}{k}\!\left[F\!\left(\dfrac{1}{k}, x\right) - E\!\left(\dfrac{1}{k}, x\right)\right], & k^2 > 1 \end{cases}$ x as in 6.

8. $\int_0^\Phi \dfrac{\cos^2 u \, du}{[1 - k^2 \sin^2 u]^{1/2}} = \left(1 - \dfrac{1}{k^2}\right) F(k, \Phi) + \dfrac{1}{k^2} E(k, \Phi), \qquad k^2 < 1.$

9. $\int_0^{\pi/2} (\sin x)^{-1/2} \, dx = \int_0^{\pi/2} (\cos y)^{-1/2} \, dy$ are elliptic with value 2.62.

10. $\int_\Phi^1 \dfrac{dx}{[(1 - x^2)(x^2 - k^2)]^{1/2}} = F\!\left((1 - k^2)^{1/2}, \sin^{-1}[(1 - \Phi^2)/(1 - k^2)]^{1/2}\right),$

$$0 < k < \Phi < 1.$$

11. $\int_0^\Phi \dfrac{dx}{[x(1 - x)(1 - k^2 x)]^{1/2}} = 2F(k, \sin^{-1}\sqrt{\Phi}) \qquad 0 < k < 1, \quad 0 < \Phi < 1.$

12. $\int_\Phi^1 \dfrac{dx}{[x(1 - x)(x - k^2)]^{1/2}} = 2F\!\left((1 - k^2)^{1/2}, \sin^{-1}[(1 - \Phi)/(1 - k)]^{1/2}\right),$

$$0 < k^2 < x < 1$$

13. $\int_0^\Phi \dfrac{dx}{[(a^2 - x^2)(b^2 - x^2)]^{1/2}} = \dfrac{1}{a} F\!\left(\dfrac{b}{a}, \dfrac{u}{b}\right), \qquad 0 < x < b < a.$

$$u = \sin^{-1}(\Phi/b)$$

minimum depending upon k. The axes of symmetry about $0, \pm\pi/2, \ldots$ imply that $E(k, \eta)$ and $F(k, \eta)$ need only be tabulated from $\eta = 0$ to $\pi/2$.

Many properties and applications of elliptic integrals are developed and tabulated by Hancock [22], Whittaker and Watson [23], Byrd and

Friedman [24], and Greenhill [25]. Table 2-2 lists a few of the more useful results.

2.8 Elliptic Functions

The Jacobi elliptic functions are defined as the inverses of the elliptic integral of the first kind. Thus if we set

$$u = u(\Phi) = \int_0^\Phi [1 - k^2 \sin^2 \theta]^{-1/2} \, d\theta$$

we are in position to define the inverse functions. Since Φ is the *amplitude* of u, we call the inverse of u by this designation and write

$$\Phi = \text{am } u \tag{2.111}$$

or

$$\Phi = \text{am}(u, \text{mod } k).$$

The quantity Φ may be regarded as angular measure so we define the trigonmetric functions of Φ and thereby generate the Jacobi elliptic functions. Thus we have

$$\sin \Phi = \sin \text{am } u = \text{sn } u$$
$$\cos \Phi = \cos \text{am } u = (1 - \text{sn}^2 u)^{1/2} = \text{cn } u$$
$$(1 - k^2 \sin^2 \Phi)^{1/2} = \text{dn } u = \Delta\Phi \tag{2.112}$$
$$\tan \Phi = \frac{\text{sn } u}{\text{cn } u} = \text{tn } u.$$

These are read, for example, "sinus amplitudinis" for sn u. The inverse functions are written

$$u = \text{sn}^{-1}(\sin \Phi, k) = \text{cn}^{-1}(\sin \Phi, k) = \text{dn}^{-1}(\Delta\Phi, k) = \text{am}^{-1}(\Phi, k)$$
$$= \text{tn}^{-1}(\tan \Phi, k).$$

Properties of these functions are given in Table 2.3.

The mathematical description of the vibration of the simple pendulum is given by the equation

$$\frac{d^2\theta}{dt^2} + \frac{g}{L} \sin \theta = 0$$

TABLE 2.3

1. $sn(0) = 0, \quad cn(0) = 1, \quad dn(0) = 1, \quad am(0) = 0$
2. $sn^2 u + cn^2 u = 1$
3. $dn^2 u - k^2 cn^2 u = 1 - k^2$
4. $k^2 sn^2 u + dn^2 u = 1$
5. $sn(-u) = -sn\,u, \quad cn(-u) = cn\,u, \quad dn(-u) = dn\,u, \quad am(-u) = -am\,u$
6. $sn(u, 0) = sin\,u, \quad cn(u, 0) = cos\,u, \quad dn(u, 0) = 1$
7. $sn(u, 1) = tanh\,u, \quad cn(u, 1) = dn(u, 1) = sech\,u$
8. $sn(u \pm 2K) = -sn\,u, \quad cn(u \pm 2K) = -cn\,u, \quad K = F(k, \pi/2)$
9. $\dfrac{d}{du}(sn\,u) = cn\,u\,dn\,u, \quad \dfrac{d}{du}(cn\,u) = -sn\,u\,dn\,u$

   $\dfrac{d}{du}(dn\,u) = -k^2 sn\,u\,cn\,u, \quad \dfrac{d}{du}(am\,u) = dn\,u$
10. $\displaystyle\int sn\,u\,du = k^{-1}\ln[dn\,u - k\,cn\,u], \quad \int cn\,u\,du = k^{-1}\cos^{-1}(dn\,u)$

    $\displaystyle\int dn\,u\,du = sin^{-1}(sn\,u), \quad \int (sn\,u)^{-1}\,du = \ln[dn\,u - cn\,u] - \ln(sn\,u)$

    $\displaystyle\int (cn\,u)^{-1}\,du = (1 - k^2)^{-1/2}\ln\left[\dfrac{dn\,u + k'\,sn\,u}{cn\,u}\right], \quad k' = (1 - k^2)^{1/2}$
11. $sn(u \pm v) = [1 - k^2 sn^2 u\,sn^2 v]^{-1}\{sn\,u\,cn\,v\,dn\,v \pm cn\,u\,sn\,v\,dn\,u\}$

where L, θ, g, and t are respectively pendulum length, angular displacement from equilibrium, acceleration of gravity, and time. A first integral is obtained by multiplying by $2(d\theta/dt)\,dt = 2d\theta$ and integrating so that

$$\frac{1}{2}\left(\frac{d\theta}{dt}\right)^2 - \frac{g}{L}\cos\theta = C \tag{2.113}$$

where C is an integration constant. Suppose $\theta = \theta_0$ is the maximum displacement from equilibrium. Then $d\theta/dt = 0$ for this value so we find $C = -(g\cos\theta_0)/L$. Solving for $d\theta/dt$ in Eq. (2.113) we have

$$dt = \left(\frac{L}{2g}\right)^{1/2}\frac{d\theta}{[\cos\theta - \cos\theta_0]^{1/2}} \tag{2.114}$$

where the plus sign† is used for the upward moving body. Our problem is now that of transforming Eq. (2.114) to standard elliptic form. Upon setting

$$k = \sin\frac{\theta_0}{2} = [(1 - \cos\theta_0)/2]^{1/2}, \qquad \cos\theta = 1 - 2k^2\sin^2\Phi \qquad (2.115)$$

we find that

$$\cos\theta - \cos\theta_0 = 2k^2\cos^2\Phi, \qquad \sin\theta = 2k\sin\Phi\,[1 - k^2\sin^2\Phi]^{1/2}$$

so that Eq. (2.114) becomes

$$dt = \left(\frac{L}{g}\right)^{1/2}\frac{d\Phi}{[1 - k^2\sin^2\Phi]^{1/2}}. \qquad (2.116)$$

Consequently the time T required for the pendulum to swing from equilibrium ($\theta = 0$) to a displacement $\theta = \theta_1$ is

$$T = \left(\frac{L}{g}\right)^{1/2}\int_0^{\Phi_0}\frac{d\Phi}{[1 - k^2\sin^2\Phi]^{1/2}} = \left(\frac{L}{g}\right)^{1/2}F(k, \Phi_0)$$

where Φ_0 is obtained from Eq. (2.115) as

$$\Phi_0 = \sin^{-1}\left[\frac{\sin(\theta_1/2)}{k}\right].$$

Let the *period P(k)* be defined as the time required to make a complete oscillation between positions of maximum displacement—that is, when $d\theta/dt = 0$. From Eq. (2.114) we have $d\theta/dt = 2k\,(L/g)^{1/2}\cos\Phi$ so $d\theta/dt = 0$ when $\Phi = \pi/2$. Therefore

$$P(k) = 4(L/g)^{1/2}\int_0^{\pi/2}\frac{d\Phi}{[1 - k^2\sin^2\Phi]^{1/2}} = 4(L/g)^{1/2}K(k)$$

where $K(k)$ is the complete elliptic integral of the first kind.

Lastly, the actual motion of the pendulum as a function of time t, can be obtained by integrating Eq. (2.116) to

$$t = \left(\frac{L}{g}\right)^{1/2}\int_0^{\Phi}\frac{dx}{[1 - k^2\sin^2 x]^{1/2}}$$

---

† If the body is moving downward $d\theta/dt$ is negative. It is easy to show that the time to move from θ_1 to θ_2 is the same as that in the reverse direction.

which in elliptic function notation is

$$\text{sn}(t(g/L)^{1/2}, k) = \sin \Phi = \frac{1}{k} \sin\left(\frac{\theta}{2}\right).$$

TABLE 2.4

1. $\dfrac{d^2y}{dx^2} = 6y^2; \; y(x) = C^2\left\{\dfrac{-k^2}{1 + k^2} + \dfrac{1}{\text{sn}^2\{C(x - x_1), k\}}\right\}, \quad C$ and x_1

   arbitrary constants, k^2 a root of $k^4 - k^2 + 1 = 0$

2. $\dfrac{d^2y}{dx^2} = Ay + By^3, \; y = C \, \text{sn}\{\lambda(x - x_0), k\}$

   λ, x_0 arbitrary constants, $k^2 = -(\lambda^2 + A)/\lambda^2, \; C^2 = -[2(\lambda^2 + A)]/B$

3. $\dfrac{d^2y}{dx^2} = A + By + Cy^2 + Dy^3;$ first integral is

   $$\left(\frac{dy}{dx}\right)^2 = a + by + cy^2 + dy^3 + ey^4$$

   Case a: $\left(\dfrac{dy}{dx}\right)^2 = h^2(y - \alpha)(y - \beta)(y - \gamma) \quad$ (cubic)

   $$y = \gamma + (\alpha - \gamma)/\text{sn}^2(-hMx, k)$$
   $$k^2 = \frac{\beta - \gamma}{\alpha - \gamma}, \quad M^2 = \frac{\alpha - \gamma}{4}$$

   Case b: $\left(\dfrac{dy}{dx}\right)^2 = h^2(y - \alpha)(y - \beta)(y - \gamma)(y - \delta) \quad$ (quartic)

   $$y = (\beta Z^2 - A\alpha)/(Z^2 - A)$$
   $$Z = \text{sn}\{hM(x - x_0), k\}, \quad A = \frac{\beta - \delta}{\alpha - \delta}$$
   $$k^2 = \frac{(\beta - \gamma)(\alpha - \delta)}{(\alpha - \gamma)(\beta - \delta)}, \quad M^2 = (\beta - \delta)(\alpha - \gamma)/4$$

4. Gambier's equation (see Davis [8, p. 183])
   $$4(y - y^2)\frac{d^2y}{dx^2} = 3(1 - 2y)\left(\frac{dy}{dx}\right)^2 + 4q(x)(y - y^2)\frac{dy}{dx}$$

   $y = 1/[1 - \Gamma^2(u)]$ where u is a solution of $u'' - q(x)u' = 0$,
   $\Gamma(u)$ is the *Weierstrass elliptic function*, solution of $(du/dx)^2 = 4u^3 - 4u$

From this form we get the $\theta = \theta(t)$ relation

$$\theta = 2 \sin^{-1}[k \, \text{sn}(t(g/L)^{1/2}, k)], \qquad k = \sin\frac{\theta_0}{2}. \qquad (2.117)$$

A short table of differential equations solvable in elliptic functions is given in Table 2.4. References to physical examples are also included.

Kamke [1] and Murphy [2] contain a collection of equations solvable in terms of elliptic functions. The swinging cord equation and the capillarity equation for two vertical plates are treated by Reddick and Miller [15]. McLachlan [26] treats the elastica as well as the hard and soft spring problem. The reduction of weakly nonlinear partial differential equations by separation has been carried out in two papers by Oplinger [27, 28]. In the first of these the nonlinearity is geometric so that tension is a function of slope. In the second the string is assumed to be viscoelastic.

In a sequence of papers Nowinski [29–31] (see also Nowinski and Woodall [32]) has utilized Galerkin's method and equations solvable in elliptic functions to examine a number of nonlinear oscillation problems. His method will be discussed in Chapter 4.

2.9 Equations with Form Homogeneity

Some equations have a homogeneity of form that admits a simplifying transformation. We examine several characteristic forms.

(a) *Homogeneity in y, y', y'', ..., $y^{(n)}$*

An equation of this class, having homogeneity of degree m, is expressible as

$$y^m f(x, y'/y, y''/y, ..., y^{(n)}/y) = 0 \qquad (2.118)$$

suggesting that a reduction of order may be possible. Thus we set

$$y = \exp\left(\int u \, dx\right) \qquad (2.119)$$

whereupon

$$y' = u e^{\int u dx}, \quad y'' = (u' + u^2) e^{\int u dx}, \quad ..., \quad y^{(n)} = F_n e^{\int u dx}$$

where F_n is a polynomial in $u, u', ..., u^{(n-1)}$. The change of variable from y to u *reduces the order* from n to $n - 1$.

As an example of this form consider the equation $(y'')^2 + (y')^2 + x^2y^2 = 0$. This is homogeneous of degree two and is rewritten as

$$y^2[(y''/y)^2 + (y'/y)^2 + x^2] = 0. \tag{2.120}$$

Setting $y = \exp(\int u \, dx)$ we find the bracketed expression becomes $(u' + u^2)^2 + u^2 + x^2 = 0$.

(b) *Homogeneity in* $y, xy', x^2y'', ..., x^ny^{(n)}$

This second class has the typical equation

$$F[y, xy', x^2y'', ..., x^ny^{(n)}] = 0 \tag{2.121}$$

which does not otherwise depend upon x. If we change the independent variable by

$$x = e^t \tag{2.122}$$

then $x \, dy/dx = dy/dt$, $\ x^2 \, dy^2/dx^2 = d^2y/dt^2 - dy/dt$, and generally

$$x^s \frac{d^sy}{dx^s} = \frac{d}{dt}\left(\frac{d}{dt} - 1\right) \cdots \left(\frac{d}{dt} - s + 1\right)y$$

where the right hand side has s factors. Thus the transformed equation has the general form

$$G\left[y, \frac{dy}{dt}, \frac{d^2y}{dt^2}, ..., \frac{d^ny}{dt^n}\right] = 0 \tag{2.123}$$

which does not specifically involve t. It is therefore a generalization of the form discussed in Section 2.6C.

2.10 Raising the Order

In Section 2.4 we discovered that the Riccati equation (2.62), that is a nonlinear equation of order one, was equivalent under the transformation equation (2.61) to the *linear* equation

$$u'' + b(x)u' + c(x)u = 0.$$

This linearization by *raising the order* can be useful in certain applications.

As a further example consider the second order equation

$$y'' + 3yy' + y^3 = f(x). \tag{2.124}$$

We can linearize this equation by means of the transformation

$$u = e^{\int y\,dx} \tag{2.125}$$

Thus

$$\frac{du}{dx} = uy$$

$$\frac{d^2u}{dx^2} = u\frac{dy}{dx} + y\frac{du}{dx} = u[y' + y^2]$$

$$\frac{d^3u}{dx^3} = u[y'' + 3yy' + y^3]$$

$$\frac{d^4u}{dx^4} = u[y''' + 4yy'' + 6y^2y' + 3(y')^2 + y^4]$$
$$\vdots$$

so that Eq. (2.124) becomes

$$\frac{d^3u}{dx^3} - uf(x) = 0. \tag{2.126}$$

If $f(x) \equiv 0$, Eq. (2.126) is immediately integrable to give

$$u = A_0x^2 + A_1x + A_2.$$

From Eq. (2.125) we have

$$y = u'/u = \frac{2A_0x + A_1}{A_0x^2 + A_1x + A_2} = \frac{2x + A}{x^2 + Ax + B} \tag{2.127}$$

where $A = A_1/A_0$, $B = A_2/A_0$ are the two arbitary constants.

2.11 A Transformation of Euler†

Euler proposed and solved the problem of reducing a class of second order equations to first order form. His method consists of replacing the

_____

† See Euler [33].

independent variable x and the dependent variable y by new variables v and u by means of the relation‡

$$x = e^{av}, \qquad y = e^v u. \qquad (2.128)$$

Here α is a constant to be determined, if possible, so that no exponential terms shall appear in the transformed equation. From Eq. (2.128) we find

$$\frac{dy}{dx} = \frac{1}{\alpha} e^{v(1-\alpha)} \left\{ \frac{du}{dv} + u \right\} \qquad (2.129)$$

$$\frac{d^2 y}{dx^2} = \frac{1}{\alpha^2} e^{v(1-2\alpha)} \left\{ \frac{d^2 u}{dv^2} + 2\frac{du}{dv} + (1 - \alpha)u \right\}. \qquad (2.130)$$

The requirement that no exponential terms appear in the transformed equation implies a certain degree of homogeneity must be present in the original equation.

Consider, as an example, the equation

$$\frac{d^2 y}{dx^2} \left(\frac{dy}{dx} \right)^{k-2} y^n = ax^m. \qquad (2.131)$$

Equation (2.131) is transformed by means of Eq. (2.128) into

$$\alpha^{-k} \exp\{[n + k - k\alpha - 1]v\} u^n \left\{ \frac{du}{dv} + u \right\}^{k-2}$$

$$\times \left\{ \frac{d^2 u}{dv^2} + 2\frac{du}{dv} + (1 - \alpha)u \right\} = ae^{mav}. \qquad (2.132)$$

The exponential terms cancel out if

$$\alpha = \frac{n + k - 1}{m + k}, \qquad m + k \neq 0.$$

With this choice of the free parameter α, Eq. (2.132) becomes

$$u^n \left\{ \frac{du}{dv} + u \right\}^{k-2} \left\{ \frac{d^2 u}{dv^2} + 2\frac{du}{dv} + \frac{m - n + 1}{m + k} u \right\} = a \left\{ \frac{n + k - 1}{m + k} \right\}^k.$$

---

‡ A simplification of this idea is much used in certain classes of linear equations (see, e.g., Morris and Brown [3]).

This equation is simpler than the original since the independent variable v is not explicitly present. Upon setting $p = du/dv$, $d^2u/dv^2 = p\,dp/du$ a reduction to first order is performed.

Several types of equations of order higher than two may be reduced to lower order by similar methods.

2.12 Equations Equivalent to Linear Equations

A few equations have already been shown to be equivalent to a linear differential equation. In this section we examine some nonlinear equations equivalent to first and second order linear equations. As early as 1902 Painlevé [34] (see Ince [6] for a survey of Painlevé's work) examined the equations

$$yy'' + a(y')^2 + f(x)yy' + g(x)y^2 = 0 \qquad (2.133)$$

and

$$y'' - (\log w)'\, y' + kq(x)y = (1 - l)y^{-1}(y')^2.$$

The solution of Eq. (2.133) is $y = u^{1/(1+a)}$† where u is a solution of the linear equation

$$u'' + f(x)u' + (a + 1)g(x)u = 0.$$

The foundation for our present discussion was laid by Pinney [35] when he considered the equation

$$y'' + p(x)y + Cy^{-3} = 0.$$

Building on this paper, Thomas [36] raised the question:
"what equations of order n have general solutions expressible as $F(u_1, u_2, \ldots, u_n)$ where $u_1, u_2, \ldots, u_n$ constitute a linearly independent set of solutions of a fixed linear differential equation?" Thomas answered the question completely when (a) the first order linear equation is inhomogeneous, (b) the linear equation is homogeneous of the second order and F depends upon only one u, and (c) the linear equation is homogeneous of the second order and F is homogeneous of degree $k \neq 0$ in two u's (i.e., $F(\lambda u, \lambda v) = \lambda^k F(u, v)$).

---

† This transformation also induces onto Eq. (2.133) a "nonlinear superposition."

A. The First Order Equation

The first of the Thomas problems is typical and illustrative of the basic concept†. Suppose $u(x)$ satisfies the equation

$$u' + p(x)u + q(x) = 0. \qquad (2.134)$$

We look for equations with solutions of the form

$$y = F(u). \qquad (2.135)$$

From this $y' = (dF/du)u'$ so that Eq. (2.134) becomes

$$y' + (pu + q)\frac{dF}{du} = 0. \qquad (2.136)$$

Now set

$$\frac{dF}{du} = f(y) \qquad (2.137)$$

so that

$$u = \int [1/f(y)] \, dy$$

whereupon Eq. (2.136) becomes

$$y' + p(x)g(y) + q(x)f(y) = 0 \qquad (2.138)$$

with

$$g(y) = f(y) \int [1/f(y)] \, dy. \qquad (2.139)$$

From Eq. (2.139) we obtain the following criteria for the integrability of Eq. (2.138) by the above method:

Any equation of the form Eq. (2.138) *satisfying either*

$$f[g/f]' = 1 \qquad (2.140a)$$

or

$$g[f/g]' = 1 \qquad (2.140b)$$

can be integrated as above.

---

† Note that these ideas can be extended to partial differential equations (see Ames [14]) but have yet to be utilized to any significant degree.

From $y = F(u)$, $F'(u) = f(y)$ we find the equation for the determination of F to be

$$F' - f(F) = 0 \qquad\qquad (2.141)$$

which, incidentally, is a special form of Eq. (2.138) with $p \equiv 0$, $q = -1$.

As an example of this consider the problem with $f(y) = y^n$, $n \neq 1$. From Eq. (2.140a) the allowable g is $g(y) = y/(1-n)$. Therefore the integrable equation, by this method, is

$$y' + p(x)y/(1 - n) + q(x)\, y^n = 0,$$

that is Bernoulli's equation.

B. THE SECOND ORDER HOMOGENEOUS EQUATION

The complete answer to Thomas' question for the second order homogeneous equation has been given by Herbst [37]. Herbst's result is only stated herein together with examples. He developed the following:

If u and v are independent solutions, with Wronskian† w(x), of the linear equation

$$Y'' - [w'(x)/w(x)]Y' + q(x)Y = 0 \qquad (2.142a)$$

then the equation

$$y'' - [w'(x)/w(x)]y' = f(y, y', w, q) \qquad (2.142b)$$

has the general solution

$$y = F(u, v) \qquad\qquad (2.142c)$$

if and only if

$$f = -q(x)Z(y) + A(y)\,(y')^2 + w^2(x)C(y) \qquad (2.142d)$$

where Z, A, and C satisfy

$$ZC' + (3 - AZ)C = 0, \qquad Z' - AZ = 1. \qquad (2.142e)$$

---

† The Wronskian is defined as $w(u, v; x) = w(x) = \begin{vmatrix} u & v \\ u' & v' \end{vmatrix}$.

F is any solution to the system

$$F_{uu} = A(F)F_u{}^2 + v^2 C(F)$$
$$F_{vv} = A(F)F_v{}^2 + u^2 C(F)$$
$$F_{uv} = A(F)F_u F_v - uvC(F) \tag{2.143}$$
$$F_u = (Z(F) - vF_v)/u.$$

The main disadvantage of this result is the difficulty in determining F. Nevertheless a number of cases can be handled by Herbst's results. We consider the following examples.

First let us find the solution to

$$y'' + p(x)y' + q(x)Z(y) = A(y)(y')^2. \tag{2.144}$$

Herbst shows that if y in Eq. (2.142c) depends only upon one solution, say u, then C must be identically zero. Thus, with Eq. (2.142e) $Z' - AZ = 1$, we search for a solution to Eqs. (2.143) of the form $y = F(u)$. Since F is independent of v, Eqs. (2.143) become

$$F'' = A(F)(F')^2 \tag{2.145}$$

since $F_u = F'$, $F_{uu} = F''$. Upon solving the linear equation $u'' + p(x)u' + q(x)u = 0$ we find $y = F(u)$ where F is the solution of Eq. (2.145).

Secondly, the solution to

$$y'' + p(x)y' + q(x)y \log y = y^{-1}(y')^2$$

is $y = e^u$, where $u'' + pu' + qu = 0$.

Thirdly, the solution to.

$$y'' - [w'/w]y' + kq(x)y = (1 - l)y^{-1}(y')^2 - k/4w^2 y^{1-4l}$$

is $y^2 = u^k v^k$ where $kl = 1$ and u, v satisfy $Y'' - (w'/w)Y' + q(x)Y = 0$.

Lastly, the solution to

$$y'' + p(x)y' + q(x)[ky - \beta y^{1-l}] = (1 - l)y^{-1}(y')^2$$

is $y = (u + \beta/k)^k$, where $kl = 1$ and $u'' + pu' + qu = 0$.

We remark, in passing, that the above four cases include as special cases fortythree of the nonlinear equations in Kamke's collection. Nowhere in this collection do these cases appear with the generality displayed herein.

The formidable problems that arise from Eqs. (2.143) have been partially circumvented by Gergen and Dressel [38]. Their result determines the exact form for F in a very general situation. We use notation analogous to that in Eqs. (2.142) and (2.143).

If in the equation

$$y'' = (w'/w)y' + f(y, y', w, q) \qquad (2.146)$$

(i) $f(y, y', w, q)$ *is differentiable in a region R for all y, $m < y < M$;*
(ii) $E(y) = f(y, 0, 0, 1) \neq 0$ *for $m < y < M$;*
(iii) *S is a domain in which $F(u_1, u_2)$ is twice differentiable and $m < F < M$;*
(iv) *$w(x)$ and $q(x)$ are arbitrary functions, $w \neq 0$;*
(v) *u_1, u_2 are linearly independent solutions of $u'' = (w'/w)u' + qu$ with Wronskian w such that u_1, u_2 are in S,*

then $y = F(u_1, u_2)$ satisfies. Eq. (2.146).
Then for $m < \eta < M$, F has the form

$$F = \Phi(\omega^{1/2}), \qquad u_1 \text{ and } u_2 \text{ in } S \qquad (2.147)$$

where $\omega(u_1, u_2)$ is a homogeneous polynomial of degree two taking positive values in S. The function Φ is the inverse function of

$$\theta(y) = \exp\left\{ \int_\eta^y E^{-1}(t)\, dt \right\}. \qquad (2.148)$$

As an example of the use of this theorem we consider the equation of Pinney [35],

$$y'' + p(x)y + Cy^{-3} = 0$$

with C constant. This equation has the form of Eq. (2.146) with $w^2 =$ constant $= C$ and $q(x) = -p(x)$. Thus

$$f(y, y', w, q) = -p(x)y - Cy^{-3}$$

so that

$$E(y) = f(y, 0, 0, 1) = y.$$

Selecting $\eta = 1$, $\theta(y) = \exp \int_1^y dt/t = y$. Thus

$$\Phi(y) = y$$

and the solution takes the form

$$F = \omega^{1/2}$$

where ω is any homogeneous polynomial of degree 2, positive in S and the solutions u_1, u_2 of

$$u'' = qu$$

have constant Wronskian w, $w^2 = C$.

As a second example the equation

$$y'' - p(x)y' - q(x)y \log y = y^{-1}(y')^2$$

will be considered. Here

$$p(x) = w'/w$$
$$f(y, y', w, q) = q(x)y \log y + (y')^2 y^{-1}$$
$$E(y) = f(y, 0, 0, 1) = y \log y.$$

To calculate Φ we evaluate

$$\theta(y) = \exp\left\{\int_{\eta}^{y} \frac{dt}{t \log t}\right\}$$

$$= \exp[\log(\log t)|_{\eta}^{y}]$$

$$= \log y, \quad for \quad \eta = e.$$

Thus $\Phi(y) = e^y$ and

$$f = \Phi(\omega^{1/2}) = \exp[\omega^{1/2}].$$

2.13 The Group Concept

Our previous efforts have been devoted to the problem of integration in the crude sense. The goal we set was the development of specific methods through which particular equations or classes of equations could be solved. At first glance the reader sees a collection of apparently disconnected methods of integration, each applicable only to one particular class of equations. This morass was substantially coordinated, in studies initiated by Lie and Klein, using the *theory of continuous groups*. The motivation behind the introduction of abstract algebraic concepts into differential equations *may* have sprung from the realization that many of the algebraic structures *are not* based upon linear hypotheses for basic operators.†

By means of continuous groups the older methods of integration are shown to depend upon one general principle. Not only does this general principle coordinate the previous methods but in addition it has proved to be a powerful instrument for further investigations.

Basic definitions for continuous groups will be given in this section. Subsequent sections will deal with applications to first and second order equations.

By a *group G* we mean a collection of elements $T_1, T_2, \ldots, T_n$, where n may be finite or infinite, having the following properties:

(i)   The group is *closed* under a *well defined operation* which we denote by $\circ$. ‡ By this we mean that $T_i \circ T_j$ is in the group for any T_i and T_j in the group.

(ii) There exists a *unique identity* which we denote by I. For each T_i in G, $T_i \circ I = I \circ T_i = T_i$.

(iii) Each element T_i in G has a *unique inverse* in G *denoted by* T_i^{-1} such that $T_i^{-1} \circ T_i = T_i \circ T_i^{-1} = I$.

(iv) $T_i \circ (T_j \circ T_k) = (T_i \circ T_j) \circ T_k$.

Basic group theory is not within the scope of this work. References to various aspects of the theory include Ledermann [39], Littlewood [40], Weyl [41], Chevally [42], and Eisenhart [43]. Additional physical applications are given by Weyl [44], Bhagavantam and Venkatarayudu [45], and Higman [46].

The groups which hold our particular interest herein are the *continuous transformation groups* (see, e.g., Eisenhart [43]), with elements that are transformations (functions), the operation being that of composition (successive transformation). For example let the finite group G_x, be composed of the following elements: $f_1(x) = x$, $f_2(x) = 1/(1-x)$, $f_3(x) = 1 - 1/x$, $f_4(x) = 1/x$, $f_5(x) = 1 - x$, $f_6(x) = x/(x-1)$. The identity is $f_1(x) = x$ since it carries any quantity into itself. As an example operation

---

† Applications of group theory in the development of similarity solutions for partial differential equations is given by Ames [14]. See also the Appendix.

‡ There is a wide range of possible operations. For example addition, multiplication, matrix multiplication, composition, etc.

we calculate $f_5 \circ f_2(x) = f_5(f_2(x)) = f_5(1/(1 - x)) = 1 - 1/(1 - x) = x/(x - 1) = f_6(x)$.

More generally let T, denoted by

$$(T): \qquad \begin{aligned} x_1 &= f(x, y) \\ y_1 &= g(x, y) \end{aligned} \qquad (2.149)$$

be a transformation carrying (x, y) into (x_1, y_1). The *inverse* transformation, that is the transformation which carries (x_1, y_1) back to its original position (x, y), exists if the Jacobian $(\partial(f, g)/\partial(x, y)) \neq 0$.† We denote this inverse by

$$(S): \qquad \begin{aligned} x &= \phi(x_1, y_1), \\ y &= \psi(x_1, y_1). \end{aligned} \qquad (2.150)$$

The result of performing T and S in succession, in either order, is the identity transformation

$$(I): \qquad \begin{aligned} x &= x_1, \\ y &= y_1. \end{aligned} \qquad (2.151)$$

We now introduce a parameter a and consider the class of transformations included in the family

$$(T_a): \qquad \begin{aligned} x_1 &= f(x, y; a), \\ y_1 &= g(x, y; a). \end{aligned} \qquad (2.152)$$

The parameter a can vary continuously over a given range. f and g are assumed to be differentiable with respect to a in that range. Any particular transformation of the family is obtained by assigning a particular value to a. We shall assume that this set forms a group and denote it by G_1 or the *group of one parameter*.‡ Examples of G_1 include

(i) The *group of translations* parallel to the x axis

$$x_1 = x + a, \qquad y_1 = y.$$

---

† This follows from the "implicit function theorem" of advanced calculus.

‡ In this volume we confine our attention to groups of one parameter in two variables although all are generalizable to r parameters and n variables (Eisenhart [43]).

(ii)   The *group of rotations* about the origin

$$x_1 = x \cos a - y \sin a, \qquad y_1 = x \sin a + y \cos a.$$

(iii)   The *group of magnifications* (contractions)

$$x_1 = a^m x, \qquad y_1 = a^n y, \qquad m, n \quad \text{real}.$$

Two basic theorems are easily proven (see Eisenhart [43]);

1.   *If a set of transformations form a group, the set of transforms of these transformations form a group.*

Let us denote by S the transformation defined by

$$x_1 = h(x, y), \qquad y_1 = k(x, y) \tag{2.153}$$

with $\partial(h, k)/\partial(x, y) \neq 0$. Let S^{-1} be its inverse. The transformation

$$\bar{T}_a = S T_a S^{-1} \tag{2.154}$$

is called the *transform* of T_a by S.

2.   *The group property of a set of transformations is invariant under a change of coordinates.*

2.14 Infinitesimal Transformations

Our next goal is to develop the heart of Lie's theory—that is the *infinitesimal transformation*. Let a_0 be that value of a corresponding to the identity transformation in Eq. (2.152). Thus

$$f(x, y; a_0) = x, \qquad g(x, y; a_0) = y.$$

Then if ε is small, the transformation

$$x_1 = f(x, y; a_0 + \varepsilon), \qquad y_1 = g(x, y; a_0 + \varepsilon) \tag{2.155}$$

is such that x_1 and y_1 differ only infinitesimally from x and y. This transformation differs only infinitesimally from I and is termed an *infinitesimal transformation*. The basic result is:

Every one parameter continuous group G_1, has one and only one infinitesimal transformation.

The proof of this result is constructive and essential for what follows. Let α be any fixed value of the parameter a and β the parameter of the corresponding inverse transformation. Thus we have

$$x_1 = f(x, y; \alpha) \qquad y_1 = g(x, y; \alpha),$$
$$x = f(x_1, y_1; \beta) \qquad y = g(x_1, y_1; \beta). \tag{2.156}$$

Let δt be small and consider

$$x' = f\{f(x, y; \alpha), \quad g(x, y; \alpha); \quad \beta + \delta t\},$$
$$y' = g\{f(x, y; \alpha), \quad g(x, y; \alpha); \quad \beta + \delta t\}. \tag{2.157}$$

With $0 < \lambda_1, \lambda_2 < 1$ an application of the mean value theorem to Eqs. (2.157) yields

$$x' = f\{f(x, y; \alpha), \quad g(x, y; \alpha); \beta\}$$

$$+ \delta t \frac{\partial}{\partial \beta} [f\{f(x, y, \alpha), \quad g(x, y; \alpha), \quad \beta + \lambda_1 \delta t\}]$$

$$= x + \xi(x, y; \alpha) \delta t, \tag{2.158}$$

$$y' = y + \eta(x, y; \alpha) \delta t. \tag{2.159}$$

The functions ξ and η are independent of δt if one neglects terms of second and higher order. Equations (2.158) and (2.159) represent an infinitesimal transformation. Geometrically, such a transformation represents a small displacement of length $[(x' - x)^2 + (y' - y)^2]^{1/2} = [\xi^2 + \eta^2]^{1/2} \delta t$ in the direction θ, defined by $\cos \theta = \xi/(\xi^2 + \eta^2)^{1/2}$, $\sin \theta = \eta/(\xi^2 + \eta^2)^{1/2}$.

Two transformation groups are *similar* when they can be derived from one another by a change of variables and parameter. We now show that *every G_1 in two variables is similar to the group of translations.* To verify this we write Eqs. (2.158) and (2.159) as

$$\delta x = \xi(x, y) \delta t, \qquad \delta y = \eta(x, y) \delta t.$$

Thus the *finite equations* of the group are found by integrating

$$\frac{dx}{\xi(x, y)} = \frac{dy}{\eta(x, y)} = dt \tag{2.160}$$

whose solutions we write as

$$F_1(x, y) = C_1, \qquad F_2(x, y) = C_2 + t.$$

with C_1, C_2 constants. Let $t = 0$ correspond to the identity transformation. Thus

$$F_1(x_1, y_1) = F_1(x, y), \qquad F_2(x_1, y_1) = F_2(x, y) + t. \qquad (2.161)$$

Now set $u = F_1(x, y)$, $v = F_2(x, y)$ so that Eq. (2.161) becomes

$$u_1 = u, \qquad v_1 = v + t, \qquad (2.162)$$

that is, the *group of translations*. This group has one and only one infinitesimal transformation, namely $\delta u = 0$, $\delta v = \delta t$.

As an example consider the group defined by

$$x_1 = a^2 x, \qquad y_1 = ay.$$

The identity corresponds to $a = 1$. The infinitesimal transformation is $x_1 = (1 + \delta t)^2 x$, $y_1 = (1 + \delta t)y$ or, to first order in δt,

$$x_1 = x + 2x \, \delta t, \qquad y_1 = y + y \, \delta t.$$

To reduce this group to the translation group it is necessary to integrate

$$dx/2x = dy/y = dt$$

whereupon Eqs. (2.161) become

$$\frac{x_1}{y_1{}^2} = \frac{x}{y^2}, \qquad \log y_1 = \log y + t.$$

The required new variables are therefore

$$u = x/y^2, \qquad v = \log y.$$

2.15 Representation of Infinitesimal Transformations

Let the given function $f(x, y)$ be subjected to the infinitesimal transformation

$$x_1 = x + \delta x = x + \xi(x, y) \, \delta t, \qquad y_1 = y + \delta x = y + \eta(x, y) \, \delta t.$$

Then

$$\begin{aligned}
\delta f(x, y) &= f(x_1, y_1) - f(x, y) \\
&= f(x + \delta x, y + \delta y) - f(x, y) \\
&= \left\{ \xi(x, y) \frac{\partial f}{\partial x} + \eta(x, y) \frac{\partial f}{\partial y} \right\} \delta t
\end{aligned} \qquad (2.163)$$

where only terms of the first order have been retained in Eq. (2.163). Conversely if $\delta f(x, y)$, that is the increment a function $f(x, y)$ assumes under an element of G_1, is known then $\xi(x, y)$ and $\eta(x, y)$ are known. *Thus the infinitesimal transformation is completely represented by the quantity*

$$Qf = \xi(x, y)\frac{\partial f}{\partial x} + \eta(x, y)\frac{\partial f}{\partial y}. \qquad (2.164)$$

In particular

$$Qx = \xi(x, y), \qquad Qy = \eta(x, y) \qquad (2.165)$$

so that we can write

$$Qf = Qx\frac{\partial f}{\partial x} + Qy\frac{\partial f}{\partial y}.$$

Examples of Eq. (2.164) include

$$-y\frac{\partial f}{\partial x} + x\frac{\partial f}{\partial y}$$

representing the *infinitesimal rotation*

$$x_1 = x - y\,\delta t, \qquad y_1 = y + x\,\delta t;$$

$$x\frac{\partial f}{\partial x} + y\frac{\partial f}{\partial y}$$

representing the *infinitesimal uniform magnification*

$$x_1 = x + x\,\delta t, \qquad y_1 = y + y\,\delta t.$$

If an element of G_1 operates on the variables x, y these are replaced by x', y' where $x' = x'(x, y)$, $y' = y'(x, y)$. Further the group property is maintained. From the chain rules we write

$$\frac{\partial f(x', y')}{\partial x} = \frac{\partial f(x', y')}{\partial x'}\frac{\partial x'}{\partial x} + \frac{\partial f(x', y')}{\partial y'}\frac{\partial y'}{\partial x}$$

$$\frac{\partial f(x', y')}{\partial y} = \frac{\partial f(x', y')}{\partial x'}\frac{\partial x'}{\partial y} + \frac{\partial f(x', y')}{\partial y'}\frac{\partial y'}{\partial y}$$

whereupon

$$
\begin{aligned}
Qf(x', y') &= \xi(x', y') \frac{\partial f}{\partial x} + \eta(x', y') \frac{\partial f}{\partial y} \\
&= \xi(x', y') \left\{ \frac{\partial f}{\partial x'} \frac{\partial x'}{\partial x} + \frac{\partial f}{\partial y'} \frac{\partial y'}{\partial x} \right\} \\
&\quad + \eta(x', y') \left\{ \frac{\partial f}{\partial x'} \frac{\partial x'}{\partial y} + \frac{\partial f}{\partial y'} \frac{\partial y'}{\partial y} \right\} \\
&= Qx' \frac{\partial f}{\partial x'} + Qy' \frac{\partial f}{\partial y'}.
\end{aligned}
$$

Let the *finite equations* of G_1, Eqs. (2.160), be

$$
\begin{aligned}
x_1 &= \phi(x, y; t) \\
y_1 &= \psi(x, y; t)
\end{aligned}
\tag{2.166}
$$

with $t = 0$ generating the identity. Any function $f(x_1, y_1)$ is therefore a function of x, y, and t. Let x, y be fixed and expand f in a Taylor series in t, thus

$$
f(x_1, y_1) = f_0 + f_0' t + \tfrac{1}{2} f_0'' t^2 + \cdots
\tag{2.167}
$$

where

$$
\begin{aligned}
f_0 &= f(x_1, y_1)\big|_{t=0} = f(x, y) \\
f_0' &= \frac{df}{dt}\bigg|_{t=0} = \frac{\partial f}{\partial x_1} \frac{dx_1}{dt} + \frac{\partial f}{\partial y_1} \frac{dy_1}{dt}\bigg|_{t=0} \\
&= \frac{\partial f}{\partial x_1} \xi(x_1, y_1) + \frac{\partial f}{\partial y_1} \eta(x_1, y_1)\bigg|_{t=0} \\
&= \frac{\partial f}{\partial x} \xi(x, y) + \frac{\partial f}{\partial y} \eta(x, y) = Qf(x, y) \\
f_0'' &= Q^2 f(x, y) \\
&\;\;\vdots \\
f_0^{(n)} &= Q^n f(x, y)
\end{aligned}
$$

where $Q^n f(x, y)$ represents the result of operating, successively, n times on f with Q (Eq. (2.164)). As a consequence we can write Eq. (2.167) in terms of the infinitesimal operator as

$$f(x_1, y_1) = f(x, y) + \sum_{n=1}^{\infty} \frac{t^n}{n!} Q^n f(x, y). \qquad (2.168)$$

Particular cases of Eq. (2.168) are illuminating. For example

$$x_1 = x + \sum_{n=1}^{\infty} \frac{t^n}{n!} Q^n x$$

$$(2.169)$$

$$y_1 = y + \sum_{n=1}^{\infty} \frac{t^n}{n!} Q^n y$$

which *are the finite equations of the group.* These equations furnish the solution of the equations

$$\frac{dx_1}{\xi(x_1, y_1)} = \frac{dy_1}{\eta(x_1, y_1)} = dt.$$

Thus the infinitesimal transformation defines the group. We therefore sometimes speak of the group Qf instead of G_1.

We now give two examples in which the finite equations (Eqs. (2.169)) are deduced from the given infinitesimal transformation.

(a)   Let $Qf = x\, \partial f/\partial x + y\, \partial f/\partial y$. Then

$$Qx = x, \qquad Q^2 x = Qx = x, ..., Q^n x = x,$$
$$Qy = y, \qquad Q^2 y = Qy = y, ..., Q^n y = y.$$

As a result

$$x_1 = x + x \sum_{n=1}^{\infty} t^n/n! = xe^t$$

$$y_1 = ye^t$$

Replacing e^t by the new parameter a we find the finite equations to be

$$x_1 = ax, \qquad y_1 = ay,$$

that is the defining equations for the group of *uniform magnifications.*

(b)   If $Qf = -y\, \partial f/\partial x + x\, \partial f/\partial y$. Then

$$Qx = -y, \qquad Q^2 x = -Qy = -x, ...$$
$$Qy = x, \qquad Q^2 y = Qx = -y, ...$$

The finite equations are therefore

$$x_1 = x\left(1 - \frac{t^2}{2!} + \frac{t^4}{4!} + \cdots\right) - y\left(t - \frac{t^3}{3!} + \cdots\right)$$

$$= x \cos t - y \sin t$$

$$y_1 = x \sin t + y \cos t$$

the defining equations for the *rotation group*.

2.16 Invariant Functions

Let the finite equations of the group be Eqs. (2.166) with the identity occurring when $t = 0$. Further let Eq. (2.164) represent the infinitesimal transformation.

A function $\theta(x, y)$ is defined to be *invariant under a given group* if, when x, y are transformed to x_1, y_1 by operations of the given group,

$$\theta(x_1, y_1) = \theta(x, y) \tag{2.170}$$

for all values of t.

The determination of the necessary and sufficient conditions for invariance follow from the expansion of $\theta(x_1, y_1)$ in powers of t (see Eq. (2.167)),

$$\theta(x_1, y_1) = \theta(x, y) + \sum_{n=1}^{\infty} \frac{t^n}{n!} Q^n \theta(x, y). \tag{2.171}$$

Since $\theta(x, y)$ is invariant under the group, the right hand side of Eq. (2.171) must be equal to $\theta(x, y)$ for all permissable values of t. For this it is *necessary and sufficient that $Q\theta$ be identically zero*, that is

$$\xi(x, y) \frac{\partial \theta}{\partial x} + \eta(x, y) \frac{\partial \theta}{\partial y} \equiv 0. \tag{2.172}$$

Thus the invariant function $Z = \theta(x, y)$ is a solution of Eq. (2.172) and $\theta(x, y) =$ constant is a solution of the equivalent Lagrange system

$$\frac{dx}{\xi} = \frac{dy}{\eta}. \tag{2.173}$$

Since Eq. (2.173) is known to have a unique solution (see, e.g., Sneddon [47]) it follows *that every G_1 in two variables has one and only one independ-*

ent invariant. Therefore, one invariant exists in terms of which all the other invariants may be expressed.

To obtain the functional invariants of a given group all we need do is integrate Eq. (2.173).

If the group is that of *rotations then*

$$Qf = -y\frac{\partial f}{\partial x} + x\frac{\partial f}{\partial y}.$$

Equation (2.173) becomes

$$\frac{dx}{-y} = \frac{dy}{x}$$

with solution $x^2 + y^2$ = constant. Thus the invariant function is $\theta(x, y) = x^2 + y^2$. *Any other invariant of the group must be a function of $x^2 + y^2$ and conversely any function of $x^2 + y^2$ is invariant under the group.*

2.17 Invariant Points, Curves, and Families of Curves

If there exists a point in the (x, y) plane for which $\xi(x, y)$ and $\eta(x, y)$ are both zero, that point is a *fixed point* under $Qf = \xi \, \partial f/\partial x + \eta \, \partial f/\partial y$. Since it is fixed under all transformations of the group we say such points are *absolutely invariant* under the group.

Suppose (x_0, y_0) is not invariant under the group. Let it be transformed by the infinitesimal transformation Qf to $(x_0 + \delta x, y_0 + \delta y)$ such that $\delta y/\delta x = \eta/\xi$. If we repeat the infinitesimal transformation indefinitely, the point P originally located at (x_0, y_0) traces out a curve which will be one of the *integral curves of*

$$dy/dx = \eta/\xi.$$

The family of integral curves $\theta(x, y)$ = constant is such that each curve is invariant under the group. However, a family of curves may also be invariant in the sense that each curve is transformed into another curve of the same family by the operations of the group. As a result the family may be invariant as a whole although each member is not invariant under the group. Let $\Omega(x, y)$ = constant be such a family of curves. If a transformation of the group carries (x, y) into (x_1, y_1) then $\Omega(x_1, y_1)$ = constant must represent the same family, for otherwise the family would not be invariant. But from the expansion Eq. (2.167), we have

$$\Omega(x_1, y_1) = \Omega(x, y) + \sum_{n=1}^{\infty} \frac{t^n}{n!} Q^n \Omega(x, y).$$

If the two families $\Omega(x, y) =$ constant, $\Omega(x_1, y_1) =$ constant are identical then

$$\sum_{n=1}^{\infty} \frac{t^n}{n!} Q^n \Omega(x, y)$$

must be constant for every fixed value of t. Thus, for every curve of the family

$$Q\Omega = \text{const.} \tag{2.174}$$

We summarize as follows:

A necessary and sufficient condition that $\Omega(x, y) =$ constant *represent a family of curves invariant (as a whole) under the group, is that* $Q\Omega =$ const. *should represent the same family of curves. That is to say* $Q\Omega$ *is a function of* Ω, *say* $F(\Omega)$.

When $F(\Omega) \equiv 0$ the individual curves are invariant (see Section 2.16). As an example we consider the rotation group with

$$Qf = -y \frac{\partial f}{\partial x} + x \frac{\partial f}{\partial y}$$

or

$$x_1 = x \cos t - y \sin t, \qquad y_1 = y \cos t + x \sin t.$$

Under this group the family of straight lines

$$y = ax$$

becomes $y_1 = bx_1$, where a and b are parameters. But

$$\frac{y_1}{x_1} = \frac{y \cos t + x \sin t}{x \sin t - y \cos t} = \frac{y}{x} + \left(1 + \frac{y^2}{x^2}\right) t + \left(\frac{y}{x} + \frac{y^3}{x^3}\right) t^2 + \cdots.$$

If the family $y/x = a$ is invariant, the family $1 + y^2/x^2 = C$ must be identical with it. This is in fact the case since a and C are related by

$$1 + a^2 = C.$$

Now with $\Omega(x, y) = y/x$ we find

$$Q\Omega = -y\frac{\partial\Omega}{\partial x} + x\frac{\partial\Omega}{\partial y} = \frac{y^2}{x^2} + 1 = \Omega^2 + 1.$$

Thus in this case $Q\Omega = F(\Omega) = \Omega^2 + 1$.

The fundamental result which forms the foundation of most of the following work is

Every invariant equation $\Omega = 0$ *is a particular integral of the equation* $Qf = 0$.

In particular, if u and v are two independent solutions of the equation

$$Qf(x, y, Z) = \xi(x, y, Z)\frac{\partial f}{\partial x} + \eta(x, y, Z)\frac{\partial f}{\partial y} + \rho(x, y, Z)\frac{\partial f}{\partial Z} = 0 \quad (2.175)$$

the most general equation, invariant under the group Qf, has the form

$$\Omega(u, v) = 0 \qquad (2.176)$$

or a special form such as $u - F(v) = 0$. Proofs of this result are available in Eisenhart [43] or Ince [6]. They follow from extension of the previous theory.

Since our applications goal for continuous groups is the solution of nonlinear differential equations we need to *extend* the group.

2.18 The Extended Group

Let the finite equations

$$x_1 = \phi(x, y; a), \qquad y_1 = \psi(x, y; a) \qquad (2.177)$$

define a G_1 in two variables. Let $p = dy/dx$ be considered as a third variable, which under the group becomes p_1. Clearly

$$p_1 = \frac{dy_1}{dx_1} = \frac{d\psi}{d\phi} = \frac{\psi_x + p\psi_y}{\phi_x + p\phi_y} = \lambda(x, y, p; a). \qquad (2.178)$$

It is easily demonstrated that the transformations

$$x_1 = \phi(x, y; a), \qquad y_1 = \psi(x, y; a), \qquad p_1 = \lambda(x, y, p; a) \qquad (2.179)$$

acting on (x, y, p) form a group which we call the *extended group* of the given group.

The given group can be further extended in an analogous way by considering the higher derivatives y'', y''', ..., $y^{(n)}$ as new variables.

Utilizing Eq. (2.165) we write the finite Eqs. (2.177) as

$$x_1 = x + t\xi(x, y) + \frac{t^2}{2!}\left[\xi\frac{\partial\xi}{\partial x} + \eta\frac{\partial\xi}{\partial y}\right] + \cdots,$$

$$y_1 = y + t\eta(x, y) + \frac{t^2}{2!}\left[\xi\frac{\partial\eta}{\partial x} + \eta\frac{\partial\eta}{\partial y}\right] + \cdots.$$

Hence, from Eq. (2.178), we find, p_1 expressible as

$$
\begin{aligned}
p_1 &= \frac{dy + t\left(\dfrac{\partial\eta}{\partial x}dx + \dfrac{\partial\eta}{\partial y}dy\right) + \cdots}{dx + t\left(\dfrac{\partial\xi}{\partial x}dx + \dfrac{\partial\xi}{\partial y}dy\right) + \cdots} \\[2mm]
&= p + t\left\{\frac{\partial\eta}{\partial x} + \left(\frac{\partial\eta}{\partial y} - \frac{\partial\xi}{\partial x}\right)p - \frac{\partial\xi}{\partial y}p^2\right\} + \cdots \\[2mm]
&= p + t\rho(x, y, p) + \cdots.
\end{aligned}
\tag{2.180}
$$

With $\rho(x, y, p)$ thus defined the infinitesimal transformation of the (first) extended group is

$$Q'f = \xi\frac{\partial f}{\partial x} + \eta\frac{\partial f}{\partial y} + \rho\frac{\partial f}{\partial p}. \tag{2.181}$$

2.19 Integration of First Order Equations

We now use the extended group and simple generalizations of Sections 2.16 and 2.17 to develop integrating factors.

Suppose

$$F(x, y, p) = 0 \tag{2.182}$$

is invariant under the extended group defined by Eq. (2.181). Thus, from Section 2.16, the necessary and sufficient condition for this invariant property is satisfied—that is, either $Q'F \equiv 0$ *per se* or because $F = 0$.

To determine and integrate the most general differential equation which is invariant under $Q'f$ we must determine two independent solutions of the partial differential equation

$$Q'f = \xi\frac{\partial f}{\partial x} + \eta\frac{\partial f}{\partial y} + \rho\frac{\partial f}{\partial p} = 0. \tag{2.183}$$

From Lagrange's analysis (see Ames [14] or Sneddon [47]) the above problem is equivalent to the determination of two distinct solutions of†

$$\frac{dx}{\xi} = \frac{dy}{\eta} = \frac{dp}{\rho}. \tag{2.184}$$

Both ξ and η are independent of p and therefore the solution $u = \alpha$ of

$$\frac{dx}{\xi} = \frac{dy}{\eta}$$

is independent of p. Let $v = \beta$ be a solution of Eq. (2.184) distinct from $u = \alpha$. It therefore follows that v must depend upon p. Then if $h(u)$ is an arbitrary function of u, $f = v - h(u)$ satisfies Eq. (2.183), that is

$$Q'\{v - h(u)\} = 0$$

so

$$v - h(u) = 0 \tag{2.185}$$

is the most general ordinary differential equation of the first order invariant under $Q'f$.

A second observation will not be verified. That is *when u is known, v can be determined by a guadrature.*

Let us now examine the differential equation

$$\frac{dx}{p(x, y)} = \frac{dy}{q(x, y)} \tag{2.186}$$

with solution $\phi(x, y) = C$. Then ϕ is an integral of the partial differential equation

$$p\frac{\partial f}{\partial x} + q\frac{\partial f}{\partial y} = 0.$$

Suppose $\phi(x, y) = C$, for at least one value of C, is not invariant under the

---

† $f \equiv$ constant is obviously a solution, albeit one of minimum interest.

group. As a family, however, the integral curves are invariant so that

$$Q\phi(x, y) = \xi\frac{\partial\phi}{\partial x} + \eta\frac{\partial\phi}{\partial y} = F(\phi)$$

where $F(\phi)$ is a function of ϕ, not identically zero. If $\Phi = \Phi(\phi)$ alone, the family of curves $\Phi = \overline{C}$ is identical to the family $\phi = C$. Let

$$\Phi = \int\frac{d\phi}{F(\phi)} \tag{2.187}$$

so that

$$Q\Phi = (Q\phi)\left(\frac{d\Phi}{d\phi}\right) = 1.$$

Φ is therefore an integral of the two equations

$$p\frac{\partial\Phi}{\partial x} + q\frac{\partial\Phi}{\partial y} = 0,$$

$$\xi\frac{\partial\Phi}{\partial x} + \eta\frac{\partial\Phi}{\partial y} = 1. \tag{2.188}$$

Solving for $\partial\Phi/\partial x$ and $\partial\Phi/\partial y$ we find

$$\frac{\partial\Phi}{\partial x} = \frac{-q}{p\eta - q\xi}, \qquad \frac{\partial\Phi}{\partial y} = \frac{p}{p\eta - q\xi}. \tag{2.189}$$

Equation (2.189) allows us to write

$$d\Phi = \frac{\partial\Phi}{\partial x}\,dx + \frac{\partial\Phi}{\partial y}\,dy = \frac{p\,dy - q\,dx}{p\eta - q\xi}. \tag{2.190}$$

Clearly $1/(p\eta - q\xi)$ *is an integrating factor for the original equation*, Eq. (2.186). The solution of $dx/p = dy/q$ is therefore

$$\int\frac{p\,dy - q\,dx}{p\eta - q\xi} = \text{const.}$$

If every individual integral curve is invariant under the group, $Q\phi$ is identically zero, that is

$$\xi\frac{\partial\phi}{\partial x} + \eta\frac{\partial\phi}{\partial y} = 0$$

whereupon $p\eta - q\xi = 0$. The infinitesimal transformation has the form

$$Qf = \varepsilon(x, y)\left[p\frac{\partial f}{\partial x} + q\frac{\partial f}{\partial y}\right]$$

and an integrating factor is not available.

2.20 Equations Invariant under Specific Groups

In this section we determine the most general differential equations which are invariant under some elementary groups. Several are examined in detail and Table 2.5 is constructed for the remainder.

(a) *Affine Group*

$$Qf = x\frac{\partial f}{\partial x}.$$

In this case $\xi = x$, $\eta = 0$, $\rho = -p$ so that the extended group is

$$Q'f = x\frac{\partial f}{\partial x} - p\frac{\partial f}{\partial p}.$$

The system

$$\frac{dx}{x} = \frac{dy}{0} = \frac{dp}{-p}$$

has solutions

$$xp = \text{const}, \qquad y = \text{const}.$$

The most general equation invariant under the group is

$$xp = F(y),$$

a separable equation.

(b) *Rotation Group*

$$Qf = -y\frac{\partial f}{\partial x} + x\frac{\partial f}{\partial y}.$$

In this case $\xi = -y$, $\eta = x$, $\rho = (1 + p^2)$ so that the extended group is

$$Q'f = -y\frac{\partial f}{\partial x} + x\frac{\partial f}{\partial y} + (1 + p^2)\frac{\partial f}{\partial p}.$$

The system

$$\frac{dx}{-y} = \frac{dy}{x} = \frac{dp}{1 + p^2}$$

has the solution

$$x^2 + y^2 = \alpha$$

TABLE 2.5

Group[a]	Invariant equation	Integrating factor
1. xf_x	$xp = F(y)$	Separable
2. yf_y	$p = yF(x)$	Separable
3. $-yf_x + xf_y$	$\dfrac{xp - y}{x + yp} = F(x^2 + y^2)$	$(x^2 + y^2)^{-1}$
4. f_x	$p = F(y)$	Separable
5. f_y	$p = F(x)$	Separable
6. $\dfrac{1}{a}f_x - \dfrac{1}{b}f_y$	$p = F(ax + by)$	Use $ax + by$ as new dependent variable (homogeneous)
7. $xf_x + yf_y$	$p = F(y/x)$	$[y - xF(y/x)]^{-1}$
8. $\dfrac{x}{a}f_x + \dfrac{y}{b}f_y$	$p = x^{((a/b)-1)}F(y^b/x^a)$	
Particular examples, $y^b = \alpha x^a$, $p = \beta x^{((a/b)-1)}$		
(a) $xf_x - yf_y$	$x\,dy = F(xy)y\,dx$	xy
(b) $2xf_x + yf_y$	$y\,dy = F(y^2/x)\,dx$	$[y^2 - xF(y^2/x)]^{-1}$
(c) $xf_x + 2yf_y$	$dy = xF(y/x^2)\,dx$	$[y - x^2F(y/x^2)]^{-1}$
9. $f_x + \dfrac{y}{x}f_y$	$p = \dfrac{y}{x} + xF(y/x)$	$1/xF(y/x)$
10. $x^2f_x + xyf_y$	$xp - y = F(y/x)$	$1/x^2F(y/x)$
11. yf_x	$\dfrac{xp - y}{p} = F(y)$	$1/y^2$
12. xf_y	$xp - y = F(x)$	$1/x^2$
13. $e^{\int \phi(x)dx}f_y$	$p - y\phi(x) = F(x)$	$e^{-\int \phi(x)dx}$

[a] Subscript notation is employed for partial derivatives to save space.

independent of p. Then the last equation can be written

$$(\alpha^2 - y^2)^{-1/2} \, dy = dp/(1 + p^2)$$

whose integral is

$$\sin^{-1}(y/\alpha) - \tan^{-1} p = \beta$$

or

$$\tan^{-1} y/x - \tan^{-1} p = \beta.$$

This may be written as

$$\frac{\dfrac{y}{x} - p}{1 + \dfrac{y}{x} p} = \tan \beta$$

using the properties of inverse trigonometric functions.

Consequently, the most general differential equation invariant under this group is

$$\frac{xp - y}{x + yp} = F(x^2 + y^2)$$

or $(x - yF) \, dy - (y + xF) \, dx = 0$ admits of the integrating factor $(x^2 + y^2)^{-1}$.

2.21 Extension to Second Order Equations

In this section we do no more than sketch the extension of the preceding remarks to second order equations and give some examples. More detail is given by Forsyth [48] and Page [49].

Let $u = \alpha, \cdot v = \beta, w = \gamma \, (\alpha, \beta, \gamma = $ constants) be distinct solutions of the system

$$\frac{dx}{\xi} = \frac{dy}{\eta} = \frac{dy'}{\rho} = \frac{dy''}{\delta} \tag{2.191}$$

with $u = u(x, y)$, $v = v(x, y, y')$, $w = w(x, y, y', y'')$. The most general differential equation invariant under the twice extended group

$$Q''f = \xi \frac{\partial f}{\partial x} + \eta \frac{\partial f}{\partial y} + \rho \frac{\partial f}{\partial y'} + \delta \frac{\partial f}{\partial y''} \tag{2.192}$$

is

$$w = F(u, v) \qquad (2.193)$$

where F is an arbitrary function of its arguments. There is no loss in generality in taking $w = dv/du$ so that the second order equation $w = F(u, v)$ is equivalent to the first order equation.

The results given in Table 2-6 are from Page [49].

TABLE 2.6

	Group	Invariant equation
1.	f_x	$y'' = F(y, y')$
2.	f_y	$y'' = F(x, y')$
3.	xf_x	$xy'' = y'F(y, xy')$
4.	yf_y	$y'' = yF(x, y'/y)$
5.	$xf_x + yf_y$	$xy'' = F(y/x, y')$
6.	$\phi(x)f_y$	$y'' = p(x)y' + q(x)y + r(x)$

REFERENCES

1. Kamke, E., "Differentialgleichungen (Lösungsmethoden und Lösungen)," Vol. I, Akad. Verlagsges., Leipzig, 1956.
2. Murphy, G. M., "Ordinary Differential Equations and Their Solutions," Van Nostrand, Princeton, New Jersey, 1960.
3. Morris, M., and Brown, O. E., "Differential Equations," Prentice–Hall, Englewood Cliffs, New Jersey, 1942.
4. Ames, W. F., *Ind. Eng. Chem.* **52**, 517 (1960).
5. Ames, W. F., *Ind. Eng. Chem. Fundamentals* **1**, 214 (1962).
6. Ince, E. L., "Ordinary Differential Equations," Dover, New York, 1958.
7. Cunningham, W. J., "Introduction to Nonlinear Analysis," McGraw–Hill, New York, 1958.
8. Davis, H. T., "Introduction to Nonlinear Differential and Integral Equations," Dover, New York, 1962.
9. Stoker, J. J., "Nonlinear Vibrations," Wiley (Interscience), New York, 1950.
10. Bickley, W. G., *Phil. Mag.* **17**, 603 (1934).
11. Chambré, P. L., *J. Chem. Phys.* **20**, 1795 (1952).
12. Carlson, A. J., Ivy, A. C., Krasno, L. R., and Andrews, A. H., *Quart. Bull. Northwestern Univ. Med. School* **16**, 254 (1942).
13. Hathaway, A. S., *Am. Math. Monthly* **28**, 93 (1921).

14. Ames W. F., "Nonlinear Partial Differential Equations in Engineering," Academic Press, New York, 1965.
15. Reddick, H. W., and Miller, F. H., "Advanced Mathematics for Engineers," 3rd ed., Wiley, New York, 1956.
16. Langmuir, I., and Blodgett, K. B., *Phys. Rev.* **22**, 347 (1923).
17. Ames, W. F., ed., Ad hoc exact techniques for nonlinear partial differential equations, *In* "Nonlinear Partial Differential Equations—Methods of Solution" Academic Press, New York, 1967.
18. Jahnke, E., Emde, F., and Loesch, F., "Tables of Functions," 6th ed. McGraw-Hill, New York, 1960.
19. Abramowitz, M., and Stegun, I. A., eds., "Handbook of Mathematical Functions," Natl. Bur. Std., U.S. Govt. Printing Office, Washington, D.C., 1964.
20. Erdelyi, A., ed., "Bateman Manuscript Project," McGraw-Hill, New York, 1955.
21. Fletcher, A., Miller, J. C. P., Rosenhead L., and Comrie, L. J., "An Index of Mathematical Tables," Addison-Wesley, Reading, Massachusetts, 1962.
22. Hancock, H., "Elliptic Integrals," Dover, New York, 1958.
23. Whittaker, E. T., and Watson, G. N., "A Course of Modern Analysis," 4th ed., Cambridge Univ. Press, London and New York, 1927.
24. Byrd, P. F., and Friedman, M. D., "Handbook of Elliptic Integrals for Physicists and Engineers," Springer, Berlin, 1954.
25. Greenhill, A. G., "Applications of Elliptic Functions," Dover, New York, 1959.
26. McLachlan, N. W., "Ordinary Nonlinear Differential Equations," 2nd ed.. Oxford Univ. Press, Londen and New York, 1958.
27. Oplinger, D. W., *J. Acoust. Soc. Am.* **32**, 1529 (1960).
28. Oplinger, D. W., *Proc. Intern. Congr. Rheol. 4th, Providence, 1963*, pt. 2, p. 231. Wiley (Interscience), New York, 1965.
29. Nowinski, J. L., *AIAA J.* **1**, 617 (1963).
30. Nowinski, J. L., *J. Appl. Mech.* **31**, 72 (1964).
31. Nowinski, J. L., Nonlinear oscillations and stability of plates under large initial stress, Tech. Rept. No. 51, Dept. of Mech. Eng., Univ. of Delaware, Newark, Delaware, October, 1965.
32. Nowinski, J. L., and Woodall, S. R., *J. Acoust, Soc. Am.* **36**, 2113 (1964).
33. Euler, L., *Commun. Acad. Petrop.* **3**, 124 (1728).
34. Painlevé, P., *Acta Math.* **25**, 1, (1902).
35. Pinney, E., *Proc. Am. Math. Soc.* **1**, 681 (1950).
36. Thomas, J. M., *Proc. Am. Math. Soc.* **3**, 899 (1952).
37. Herbst, R. T., *Proc. Am. Math. Soc.* **7**, 95 (1956).
38. Gergen, J. J., and Dressel, F. G., *Proc. Am. Math. Soc.* **16**, 767 (1965).
39. Ledermann, W., "Introduction to the Theory of Finite Groups," 2nd ed. Wiley (Interscience), New York, 1953.
40. Littlewood, D. E., "The Theory of Group Characters," Oxford Univ. Press, London and New York, 1952.
41. Weyl, H., "The Classical Groups," Princeton Univ. Press, Princeton, New Jersey, 1939.

42. Chevally, C., "The Theory of Lie Groups," Vol. 1, Princeton Univ. Press, Princeton, New Jersey, 1946.
43. Eisenhart, L. P., "Continuous Groups of Transformations," Dover, New York, 1961.
44. Weyl, H., "The Theory of Groups and Quantum Mechanics," Dover, New York, 1950.
45. Bhagavantham, S., and Venkatarayudu, T., "Theory of Groups and Its Application to Physical Problems," Andhra Univ., Waltair, India, 1948.
46. Higman, P., "Applied Group Theoretic Methods," Oxford Univ. Press, London and New York, 1958.
47. Sneddon, I. N., "Elements of Partial Differential Equations," McGraw-Hill, New York, 1957.
48. Forsyth, A. R., "Theory of Differential Equations," Vols. 1–4, Dover, New York, 1959.
49. Page, J. M., "Ordinary Differential Equations, with an Introduction to Lie's Theory of Groups of One Parameter," London, 1897.

3

EXAMPLES FROM TRANSPORT
PHENOMENA

Introduction

The physical orientation of this volume is toward *transport phenomena.*
This general classification is herein interpreted as consisting of the four
basic areas of diffusion, reaction, heat transfer and fluid flow. Consequently, the examples and special techniques of this chapter are drawn
from these disciplines. Some of the problems ultimately require approximate or numerical solution. These topics are the subject of Chapters 4 and 5.

3.1 Matrices and Chemical Reactions

A problem of importance in reaction kinetics is that of evaluating rate
constants and their ratios for the describing systems of differential equations. These equations are generally nonlinear. Nevertheless a simpler
canonical form can often be developed by means of a matrix method. The
motivation was the application of matrix theory in the determination of
normal coordinates for the vibrations of linear elastic systems with many
degrees of freedom. The normal coordinate process is based upon the
diagonalization of the system coefficient matrix (see for example Tong
[1] or Pipes [2]). In the strict sense this application of matrix algebra is
useful only for linear systems. However, the essentials of that procedure
can be employed on some systems of nonlinear equations providing
precautions are taken. The ideas and examples given herein are due to
Ames [3].

Several elementary concepts from matrix algebra are required in the sequel. The interested reader may find the pertinent details in the work of Pipes [4], Albert [5], or Hildebrand [6]. Three types of elementary operations or transformations upon the rows of a matrix can be defined: (a) The interchange of two rows; (b) The multiplication of the elements of any row by the same nonzero number α; (c) The addition (or subtraction) to any row of a multiple of any other row. While it is possible to define corresponding column operations the nonlinearity of the equations preclude their use.

In addition to the foregoing operations two theorems are important for the development of the canonical form.

I. *If A is a nonsingular† square matrix, then A can be transformed into the identity matrix by elementary row transformations alone.*

II. *Every nonzero m by n matrix A whose rank‡ is r(r $\neq$ 0) can be transformed by elementary row transformations alone into a matrix of the form*

$$\begin{bmatrix} I_r & 0 \\ 0 & 0 \end{bmatrix} \tag{3.1}$$

where I_r is the r by r identity matrix and 0 is some zero matrix.

The application of Theorem II is fundamental to the development of the canonical form because the matrix of the kinetic equations will invariably be a non-square matrix initially. In illustration of this procedure we use a reaction system similar to Eqs. (1.22). These are arranged in matrix form as follows: Along the top of the matrix place the individual terms. Along the side of the array place the derivative to be considered. At those positions where the differential equation has a term insert the appropriate

---

† The square matrix A is said to be nonsingular if the determinant of A is not zero.
‡ The matrix A has rank r if and only if it has a nonvanishing determinant of order r and none of order greater than r.

coefficient. Proceeding thusly we obtain the matrix form

$$
\begin{array}{cccc}
x_1x_2, & x_2x_3, & x_2x_5, & x_2x_6 \\
\end{array}
\qquad \frac{d(\ )}{dt}
$$

$$
\begin{bmatrix}
-k_1 & 0 & 0 & 0 \\
-k_1 & -k_2 & -k_3 & -k_4 \\
k_1 & -k_2 & 0 & 0 \\
0 & k_2 & 0 & 0 \\
0 & 0 & -k_3 & 0 \\
0 & 0 & k_3 & -k_4 \\
0 & 0 & 0 & k_4
\end{bmatrix}
=
\begin{bmatrix}
x_1 \\
x_2 \\
x_3 \\
x_4 \\
x_5 \\
x_6 \\
x_7
\end{bmatrix}. \qquad (3.2)
$$

From Benson [7] we observe that Eq. (3.2) must have redundancies. These result since Eq. (3.2) describes four reactions. Therefore only four differential equations are required.

No special treatment is needed to pick out the redundancies† as shall be seen in the development of the canonical form. However, useful information for the development of the canonical form is available from the selection of products of reaction which do not themselves react. We call these *dead end* variables. The dead and variables of this reaction system are clearly x_4 and x_7.

Keeping in mind that elimination of the dead end variables is usually expedient the matrix equation (3.2) is now reduced to diagonal form by application of the, previously described, elementary matrix operations *on the rows alone*. Of course the same operations are also carried out on the right hand column vector. In the event that any row is reducible to zero a

---

† The form of redundancies is not unique.

redundancy has been found and no further consideration of this row is necessary. Further, by Theorem II, with the exception of the redundant rows, the remainder of the matrix (4 × 4 in this case) always permits of reduction to the identity matrix. However, we often find it convenient to interrupt the reduction at the diagonal form.

A variety of reduction schemes are possible but some systematic process is preferred. To this end arrange the system so that the element in the upper left hand corner *is not zero*. This can always be done by rearranging the rows. All elements in the first column are now eliminated by adding (or subtracting) a suitable multiple of the first row to every other row which has a nonzero element in the first column. Thus for our Eq. (3.2) there results

$$
\begin{array}{ccccc}
x_1 x_2, & x_2 x_3, & x_2 x_5, & x_2 x_6 & \dfrac{d(\;)}{dt}
\end{array}
$$

$$
\begin{bmatrix}
-k_1 & 0 & 0 & 0 \\
0 & -k_2 & -k_3 & -k_4 \\
0 & -k_2 & 0 & 0 \\
0 & k_2 & 0 & 0 \\
0 & 0 & -k_3 & 0 \\
0 & 0 & k_3 & -k_4 \\
0 & 0 & 0 & k_4
\end{bmatrix}
=
\begin{bmatrix}
x_1 \\
x_2 - x_1 \\
x_1 + x_3 \\
x_4 \\
x_5 \\
x_6 \\
x_7
\end{bmatrix}.
\qquad (3.3)
$$

Now consider the second column, second row position, usually designated as the (2,2) position. This row has three elements. In this problem, recalling that x_4, x_7 are dead end, interchange the third and second row, thus utilizing the fact that the third row (for $x_3 + x_1$) contains only one element. Further, exchange of the new third row with the sixth, and the fourth with the fifth gives the new system

$$
\begin{array}{cccc}
x_1x_2, & x_2x_3, & x_2x_5, & x_2x_6
\end{array}
\qquad \dfrac{d(\ )}{dt}
$$

$$
\begin{bmatrix}
-k_1 & 0 & 0 & 0 \\
0 & -k_2 & 0 & 0 \\
0 & 0 & k_3 & -k_4 \\
0 & 0 & -k_3 & 0 \\
0 & k_2 & 0 & 0 \\
0 & -k_2 & -k_3 & -k_4 \\
0 & 0 & 0 & k_4
\end{bmatrix}
=
\begin{bmatrix}
x_1 \\
x_1 + x_3 \\
x_6 \\
x_5 \\
x_4 \\
x_2 - x_1 \\
x_7
\end{bmatrix}. \qquad (3.4)
$$

Using the new element in the (2, 2) position of Eq. (3.4) all other elements in the second column can be eliminated by elementary transformations. Continuing this procedure the final reduction becomes

$$
\begin{array}{cccc}
x_1x_2, & x_2x_3, & x_2x_5, & x_2x_6
\end{array}
\qquad \dfrac{d(\ )}{dt}
$$

$$
\begin{bmatrix}
-k_1 & 0 & 0 & 0 \\
0 & -k_2 & 0 & 0 \\
0 & 0 & -k_3 & 0 \\
0 & 0 & 0 & -k_4 \\
0 & 0 & 0 & 0 \\
0 & 0 & 0 & 0 \\
0 & 0 & 0 & 0
\end{bmatrix}
=
\begin{bmatrix}
x_1 \\
x_1 + x_3 \\
x_5 \\
x_6 + x_5 \\
x_4 + x_3 + x_1 \\
x_2 - 2x_1 - 2x_5 - x_6 - x_3 \\
x_7 + x_6 + x_5
\end{bmatrix}. \qquad (3.5)
$$

The four by four diagonal matrix in Eq. (3.5) can be further reduced to the identity matrix I_4 by division of each row by $-k_i$. This may or may not be helpful. From the last three rows of Eq. (3.5) the redundancies are easily seen to be

$$\frac{d(x_7 + x_6 + x_5)}{dt} = 0$$

or

$$x_7 = c_7 - x_6 - x_5. \tag{3.6}$$

Similarily,

$$x_4 = c_4 - x_3 - x_1 \tag{3.7}$$

$$\begin{aligned} x_2 &= 2x_1 + x_3 + 2x_5 + x_6 + c_2 \\ &= x_1 + c_4 - x_4 + x_5 + c_7 - x_7 + c_2 \\ &= x_1 - x_4 + x_5 - x_7 + c_4 + c_7 + c_2. \end{aligned} \tag{3.8}$$

In Eqs. (3.6, 3.7, and 3.8) the c's are combinations of the initial conditions, for example

$$c_2 = x_2(0) - 2x_1(0) - x_3(0) - 2x_5(0) - x_6(0)$$

where it is understood that some of these initial values may be zero. Eqs. (3.6, 3.7, and 3.8) are the required expressions for x_7, x_4, and x_2 in terms of the remaining variables of the system. Similar relations could be inferred from the stoichiometry of the system.

Upon elimination of the redundancies from Eq. (3.5) there results

$$
\begin{array}{cccc}
x_1 x_2, & x_2 x_3, & x_2 x_5, & x_2 x_6 & \frac{d(\ )}{dt}
\end{array}
$$

$$
\begin{bmatrix}
-k_1 & 0 & 0 & 0 \\
0 & -k_2 & 0 & 0 \\
0 & 0 & -k_3 & 0 \\
0 & 0 & 0 & -k_4
\end{bmatrix}
=
\begin{bmatrix}
x_1 \\
x_1 + x_3 \\
x_5 \\
x_5 + x_6
\end{bmatrix}
\tag{3.9}
$$

which shall be termed the *canonical form* of the system. In differential equation form this reads

$$\frac{dx_1}{dt} = -k_1 x_1 x_2 \tag{3.10a}$$

$$\frac{d(x_1 + x_3)}{dt} = -k_2 x_2 x_3 \tag{3.10b}$$

$$\frac{dx_5}{dt} = -k_3 x_2 x_5 \tag{3.10c}$$

$$\frac{d(x_5 + x_6)}{dt} = -k_4 x_2 x_6 \tag{3.10d}$$

where x_2 is expressible in terms of x_1, x_3, x_5, x_6 by means of Eq. (3.8).

The canonical set, Eqs. (3.10), does not have all the advantages of a corresponding linear system. However, there are multiple advantages for proceeding with the analysis via the canonical form instead of the unreduced system.

The *first advantage* is the ease with which the rate constant ratios can be obtained from the data. To illustrate this approach divide Eq. (3.10b) by Eq. (3.10a) to obtain

$$\frac{dx_3}{dx_1} = \frac{k_2}{k_1} \frac{x_3}{x_1} - 1. \tag{3.11}$$

Division of Eq. (3.10d) by Eq. (3.10c) yields an equation of the same form as Eq. (3.11) namely

$$\frac{dx_6}{dx_5} = \frac{k_4}{k_3} \frac{x_6}{x_5} - 1. \tag{3.12}$$

Both Eqs. (3.11) and (3.12) are simple linear first order ordinary differential equations involving the ratios $\alpha = k_2/k_1$ and $\beta = k_4/k_3$. Their solutions are the implicit functions of α and β

$$\alpha - 1 + \frac{x_1(0)\,[x_1/x_1(0)]^{\alpha} - x_1}{x_3 - x_3(0)} = 0 \tag{3.13}$$

and

$$\beta - 1 + \frac{x_5(0)\,[x_5/x_5(0)]^{\beta} - x_5}{x_6 - x_6(0)} = 0. \tag{3.14}$$

Yet a third ratio is required and it can be supplied by an examination of Eqs. (3.10a) and (3.10c) which may be rewritten in the form

$$-\frac{1}{k_3 x_5}\frac{dx_5}{dt} = x_2 = -\frac{1}{k_1 x_1}\frac{dx_1}{dt}.$$ (3.15)

This is equivalent to the new equation

$$\gamma\frac{dx_5}{x_5} = \frac{dx_1}{x_1}$$ (3.16)

with solution

$$\frac{k_1}{k_3} = \gamma = \frac{\ln x_1 - \ln x_1(0)}{\ln x_5 - \ln x_5(0)}.$$ (3.17)

Equations (3.13) and (3.14) are clearly implicit relations for α and β which require numerical solutions from observed values of x_1, x_3, x_5, and x_6. Equation (3.17) is an explicit relation for γ.

A *second advantage* accrues to this technique from a computational vantage. If the canonical equations are used for computation, after the rate constants ratios have been determined, the number of parameters is reduced from 4 to 1, for setting of one rate constant fixes the remaining three through the known ratios. In addition the amount of computation is generally reduced.

Further examples are detailed by Ames [3].

3.2 Kinetics and the Z Transform

One analytic method having considerable utility in the solution of linear stagewise processes is the Z transform. † Properties and applications of this transform to certain chemical engineering problems are discussed in an article by Murdoch [8]. Abraham [9] applies the technique to integrate the infinite set of linear kinetic equations characterizing linear polymers. The Z transform is widely used, sometimes under a different name, in the analysis of sampled-data control systems. Consequently, standard

---

† Also called the "finite difference transform" and the "power series transform" by several authors. The theory of the "generating function" in probability is closely related to these concepts.

texts in that discipline provide tables and operational properties of the transform (see, e.g., Jury [10] or Ragazzini and Franklin [11]). Texts concerned with the Laurent series in complex variable, such as Carrier *et al.* [12] or Churchill [13] are helpful. Bharucha–Reid [14] presents a variety of generating function applications to problems of stochastic processes. His monograph is helpful in both formulation and solution of problems of polymerization kinetics. In the sequel we first give some pertinent definitions and properties of Z transforms that pertain to the *nonlinear problems* of polymer kinetics.

The Z transform of a distribution of polymer chain lengths $\{l_n\}$ will be defined as†

$$P(z) = \sum_{n=1}^{\infty} P_n z^{-n}, \qquad |z| \geq 1. \tag{3.18}$$

Here P_n is the concentration of chains of length n (moles per unit volume) and z is an arbitrary transform variable. The following properties of the transform will be employed:

(a) $\lim_{z \to 1} P(z) = P(1) = \hat{P} = \sum_{n=1}^{\infty} P_n$ (3.19)

and convergence is assumed.

(b) If $P_n = P_n(t)$, then $P = P(z, t)$ and

$$\frac{\partial^k P}{\partial t^k} = \sum_{n=1}^{\infty} \frac{d^k P_n}{dt^k} z^{-n}, \tag{3.20}$$

$$\frac{\partial^k P}{\partial (\ln z^{-1})^k} = \sum_{n=1}^{\infty} n^k P_n z^{-n} = M_z^{(k)}. \tag{3.21}$$

(c) The Kth moment $M^{(k)}$ of the distribution is defined as

$$M^{(k)} = \sum_{n=1}^{\infty} n^k P_n \tag{3.22}$$

so that

$$M^{(k)} = \lim_{z \to 1} M_z^{(k)} \tag{3.23}$$

---

† Various modifications are possible. The term P_0 is zero herein because the physical interpretation demands that it take that value.

(d) The Z transform has a convolution theorem analogous to that for the Laplace transform. Let $\{l_n\}$ and $\{l_n^*\}$ be independent. Then

$$P(z) \cdot P^*(z) = \left(\sum_1^\infty P_n z^{-n} \right) \left(\sum_1^\infty P_n^* \, z^{-n} \right)$$

$$= \sum_{n=0}^\infty z^{-n} \left(\sum_{m=0}^n P_m \cdot P_{n-m}^* \right)$$

$$= \sum_{n=1}^\infty z^{-n} \left(\sum_{m=1}^n P_m \cdot P_{n-m}^* \right) \qquad (3.24)$$

since l_0 does not exist.

(e) The normalized Z transform is defined as

$$C(z, t) = \frac{P(z, t)}{P(1, t)} \qquad (3.25)$$

so that

$$C(z, t) = \sum_{n=1}^\infty C_n z^{-n}; \qquad C(1, t) = 1$$

where $C_n = P_n / \sum_{i=1}^\infty P_i$. For convenience we sometimes write $C_z = C(z, t)$ and designate

$$m_z^{(k)} = \frac{\partial^k C}{\partial (\ln z^{-1})^k}. \qquad (3.26)$$

From Eq. (3.25) it follows that

$$\lim_{z \to 1} m_z^{(k)} = m^{(k)} = M^{(k)}/M^{(0)} = M^{(k)}/P(1).$$

Kilkson [15] has demonstrated that the Z transform is a powerful mathematical tool for the solution of path-dependent polymerization problems. The convolution property extends its applicability to second order "random" condensations. This is a significant step since such systems are practically very common but mathematically very difficult. Kilkson [15] details several polymerizations. We describe the irreversible condensation in a batch reactor in the sequel. The reader is referred to the basic reference for additional examples.

Let A end B designate active end groups capable of reacting to form a link (AB) and let X designate an inert end group. The considered polymerization will be a condensation of bifunctional molecules of type $A - (BA)_{n-1} - B$ in the presence of monofunctional molecules or chain stoppers $A - (BA)_{n-1} - X$. If the reaction between A and B groups occurs irreversibly, according to second order kinetics, with a rate constant k independent of chain length then the overall decrease in concentration of end groups is given by

$$-\frac{d(A)}{dt} = -\frac{d(B)}{dt} = k(A)(B). \tag{3.27}$$

Here t is time and (A) represents the concentration of A.

If we designate a bifunctional molecule with n AB units by l_n and its monofunctional counterpart $(AB)_{n-1} - AX$ by l_{nx} then the condensation reaction is represented by the equations

$$l_m + l_{n-m} \xrightarrow{k} l_n$$
$$l_{mx} + l_{n-m} \xrightarrow{k} l_{nx}, \tag{3.28}$$

that is to say that an individual species may be formed from any pair of bifunctional molecules whose indices add up to n and destroyed by reaction with any molecule, regardless of length or kind. Thus the rate equation for species l_n is

$$\frac{dP_n}{d\tau} = \sum_{m=1}^{n-1} P_m P_{n-m} - P_n \left[2 \sum_{i=1}^{\infty} P_i + \sum_{i=1}^{\infty} P_{ix} \right] \tag{3.29}$$

where $\tau = kt$, $P_n(\tau)$ = concentration of l_n and $P_{nx}(\tau)$ = concentration of l_{nx}.

The species l_{nx} is created by indicial addition of a bifunctional molecule and a monofunctional one and destroyed by reaction with any bifunctional molecule. Thus

$$\frac{dP_{nx}}{d\tau} = \sum_{m=1}^{n-1} P_{mx} P_{n-m} - P_{nx} \sum_{i=1}^{\infty} P_i \tag{3.30}$$

is the rate equation for P_{nx}.

Equations (3.29) and (3.30) define the infinite sets of nonlinear rate equations describing the time dependency of the two distributions $\{l_n\}$

and $\{l_{nx}\}$. We proceed to solve these utilizing the convolution property of the Z transform. In this direction multiply Eq. (3.29) by z^{-n} and sum over n so that

$$\sum_{n=1}^{\infty} z^{-n} \frac{dP_n}{d\tau} = \sum_{n=1}^{\infty} \left[\sum_{m=1}^{n-1} P_m z^{-m} \cdot P_{n-m} z^{-(n-m)} \right]$$

$$- \sum_{n=1}^{\infty} P_n z^{-n} \left[2 \sum_{i=1}^{\infty} P_i + \sum_{i=1}^{\infty} P_{ix} \right]. \qquad (3.31)$$

From the Z transform properties given in Eqs. (3.19)–(3.24), in terms of $P(z, \tau)$, Eq. (3.31) becomes

$$\frac{\partial P(z, \tau)}{\partial \tau} = [P(z, \tau)]^2 - P(z, \tau)[2P(1, \tau) + P_x(1, \tau)]. \qquad (3.32)$$

Transformation of Eq. (3.30) in a similar manner yields, for the distribution l_{nx}

$$\frac{\partial P_x(z, \tau)}{\partial \tau} = P(z, \tau)P_x(z, \tau) - P_x(z, \tau)P(1, \tau). \qquad (3.33)$$

By application of the Z transform the two infinite sets of ordinary differential equations have been replaced by the two modest, but nonlinear, partial differential equations (3.32) and (3.33). We now examine these equations and develop their solutions.

First if we set $z = 1$ in Eqs. (3.32) and (3.33) there results

$$\frac{dP(1, \tau)}{d\tau} = - [P(1, \tau)]^2 - P(1, \tau)P_x(1, \tau) \qquad (3.34)$$

$$\frac{dP_x(1, \tau)}{d\tau} = 0, \qquad (3.35)$$

that is the rate equations for the total number of molecules of each kind. From these we observe that the total number of monofunctional molecules may not change but the number of bifunctional molecules decreases as a result of internal condensation and reaction with monofunctional ones.

The solution of Eqs. (3.32) and (3.33) is simplified by transformation to the normalized transforms $C(z, \tau)$ and $C_x(z, \tau)$ defined via Eq. (3.25).

From Eq. (3.25) we have by τ differentiation

$$\frac{\partial C}{\partial \tau} = \frac{1}{P(1, \tau)}\frac{\partial P(z, \tau)}{\partial \tau} - \frac{C}{P(1, \tau)}\frac{dP(1, \tau)}{d\tau}. \qquad (3.36)$$

Substituting Eqs. (3.32) and (3.34) into Eq. (3.36) yields

$$\frac{\partial C}{\partial \tau} = P(1, \tau)C[C - 1]. \qquad (3.37)$$

The time dependency for normalized transform $C_x = P_x(z, \tau)/P_x(1, \tau)$ is obtained in a similar manner to Eq. (3.37) as

$$\frac{\partial C_x}{\partial \tau} = P(1, \tau)C_x[C - 1]. \qquad (3.38)$$

Upon division of Eq. (3.37) by Eq. (3.38) and integrating we observe that

$$\frac{C}{C_0} = \frac{C_x}{C_{x0}} \qquad (3.39)$$

where C_0 and C_{x0} represent feed distributions. Thus we see that the distributions are simply related for dissimilar feeds and are identical for identical feed distributions.

Integration is further facilitated by setting

$$f = P(1, \tau)/P(1, 0)$$
$$= \text{fraction of original } B\text{-type ends remaining} \qquad (3.40)$$

and

$$X_0 = P_x(1, 0)/P(1, 0)$$
$$= \text{ratio of chain stoppers to bifunctional molecules at } \tau = 0. \qquad (3.41)$$

Upon taking the τ derivative of Eq. (3.40) and substituting for $P(1, \tau)$ and $P_x(1, \tau)$ we find the equation for f

$$\frac{1}{P(1, 0)}\frac{df}{d\tau} = -f[f + X_0]. \qquad (3.42)$$

An expression for $C(z, \tau)$ in terms of C_0, f, and X_0 is obtainable by dividing

Eq. (3.36) by Eq. (3.42) and integrating. Thus

$$\frac{C}{C-1} = \left[\frac{C_0}{C_0 - 1}\right]\left[\frac{f + X_0}{1 + X_0}\right]. \tag{3.43}$$

For simplicity in notation set

$$1 - q = (f + X_0)/(1 + X_0) \qquad \text{or} \qquad q = (1 - f)/(1 + X_0). \tag{3.44}$$

From Eq. (3.44) it is clear that $0 \leq q \leq (1 + X_0)^{-1}$ and $\lim_{X_0 \to 0} q = 1 - f$. In terms of $q = q(f, X_0)$ we can rewrite Eq. (3.43) as

$$C = (1 - q)C_0/(1 - qC_0) \tag{3.45}$$

and Eq. (3.39) becomes

$$C_x = (1 - q)C_{x0}/(1 - qC_{x0}). \tag{3.46}$$

To establish the time dependency of q we note that Eq. (3.42) is an integrable form of the Riccati equation (see Chapter 2) whose integral is

$$f = X_0[(1 + X_0)\exp\{X_0 \cdot P(1, 0)\tau\} - 1]^{-1}. \tag{3.47}$$

When $X_0 = 0$, Eq. (3.42) has the solution

$$f = [1 + P(1, 0)\tau]^{-1}. \tag{3.47a}$$

Equations (3.44)–(3.47) constitute the solution to the problem for the normalized Z transforms C and C_x. There remains a determination of C_0. If, for example, the feed is all monomer then $P_1(0) = 1$, $P_j(0) = 0$, $j > 1$. Thus $C_0 = z^{-1}$. From Eq. (3.45) we then have

$$C = (1 - q)z^{-1}/(1 - qz^{-1})$$

$$= (1 - q)\sum_{n=1}^{\infty} q^{n-1}z^{-n} \tag{3.48}$$

and therefore

$$C_n = P_n / \sum_{i=1}^{\infty} P_i = (1 - q)q^{n-1}$$

a result known to Flory [16] as the "most probable distribution." Further discussion of this example and other problems solvable in a similar manner may be found in the work of Kilkson [15].

Approximate methods utilizing transforms are found in Chapter 4.

3.3 Kinetics and Heat Transfer

If Q is the monomolecular heat of reaction, k the thermal conductivity, and W the reaction velocity then the appropriate equation for the thermal balance between the heat generated by a chemical reaction and that conducted away is

$$k \nabla^2 T = -QW. \tag{3.49}$$

The usual expression for W is the Arrhenius relation

$$W = ca \exp[-E/RT] \tag{3.50}$$

where c, a, E, R, and T are concentration, frequency factor, energy of activation, gas constant, and temperature respectively.

As a first approximation, in the case of small temperature ranges, write

$$-\frac{E}{RT} = -\frac{E}{RT_0}\left[\frac{1}{1 + (T - T_0)/T_0}\right]$$

$$= -\frac{E}{RT_0}\left[1 - \frac{T - T_0}{T_0} + \left(\frac{T - T_0}{T_0}\right)^2 - \cdots\right]$$

$$\approx -\frac{E}{RT_0}\left[1 - \frac{T - T_0}{T_0}\right]. \tag{3.51}$$

Employing the approximation of Eq. (3.51) in Eq. (3.50) we find the equation for dimensionless temperature, $\theta = (E/RT_0^2)(T - T_0)$, to be

$$\nabla^2\theta + A \exp \theta = 0 \tag{3.52}$$

where

$$A = (QEca/kRT_0^2) \exp\left(-\frac{E}{RT_0}\right)$$

and ∇^2 represents the Laplace operator (see, e.g., Moon and Spencer [17] for an extensive tabulation of this and other linear operators).

Equation (3.52) also occurs in certain problems of two dimensional vortex motion of incompressible fluids as given by Bateman [18], in the theory of the space charge of electricity around a glowing wire (Richardson [19]), and in the nebular theory for the distribution of mass of gaseous interstellar material under the influence of its own gravitational field. Walker [20] considers this problem in detail utilizing general solutions.

In the following we concentrate attention upon the one dimensional problems obtained by using cartesian, circular cylindrical, and spherical coordinates. In these cases Eq. (3.52) becomes (see Moon and Spencer [17])

$$\frac{d^2\theta}{dx^2} + \frac{n}{x}\frac{d\theta}{dx} = -A \exp \theta \tag{3.53}$$

with boundary conditions

$$\frac{d\theta}{dx} = 0 \quad \text{at} \quad x = 0$$
$$\theta = 0 \quad \text{at} \quad x = 1. \tag{3.54}$$

Setting $n = 0, 1, 2$ generates the appropriate equation for cartesian, circular cylindrical or spherical coordinates, respectively. The cases $n = 0$ and $n = 1$ have been exactly solved at this writing (1967) and these are considered herein.

For the case $n = 0$ Eq. (3.53) becomes

$$\frac{d^2\theta}{dx^2} = -A \exp \theta \tag{3.55}$$

an equation of type (2.91). Its integration is readily accomplished by the methods of Chapter 2.

For $n = 1$, Chambré [21] develops an exact solution in an ingenious manner. Equation (3.53), with $n = 1$, becomes

$$\frac{d^2\theta}{dx^2} + \frac{1}{x}\frac{d\theta}{dx} = -A \exp \theta \tag{3.56}$$

which is easily rewritten as

$$\frac{d}{dx}\left[x\frac{d\theta}{dx}\right] = -Ax \exp \theta. \tag{3.56a}$$

We let p be arbitrary, for the moment, and attempt the combined transformation

$$\omega = x\frac{d\theta}{dx}, \quad u = x^p e^\theta. \tag{3.57}$$

Thus

$$\frac{d\omega}{dx} = \frac{d\omega}{du}\frac{du}{dx}$$

$$= \frac{d\omega}{du}\left[px^{p-1}e^{\theta} + x^{p}e^{\theta}\frac{d\theta}{dx}\right]$$

$$= u\frac{d\omega}{du}\left[px^{-1} + \frac{d\theta}{dx}\right]. \qquad (3.58)$$

Multiplying by x and employing Eq. (3.57) we can finally write Eq. (3.58) as

$$x\frac{d\omega}{dx} = u\frac{d\omega}{du}[p + \omega]$$

and because of Eq. (3.56a)

$$-Ax^2e^{\theta} = u\frac{d\omega}{du}[p + \omega]. \qquad (3.59)$$

The integration of Eq. (3.59) can be completed if its left hand side is a function of u and ω. In particular if $p = 2$ the left hand side is $-Au$. Thus, with $p = 2$, Eq. (3.59) becomes

$$\frac{d\omega}{du} = -\frac{A}{2 + \omega} \qquad (3.60)$$

for u not identically zero. This separable first order equation integrates immediately to

$$\omega^2 + 4\omega = -2Au + D. \qquad (3.61)$$

To evaluate the integration constant D we apply the boundary condition at $x = 0$. Here $u = 0$ and since $\omega = x\, d\theta/dx$, $\omega = 0$. Therefore $D = 0$ and the first integral becomes

$$\omega^2 + 4\omega = -2Au$$

or in terms of the original variables

$$x^2\left(\frac{d\theta}{dx}\right)^2 + 4x\frac{d\theta}{dx} = -2Ax^2e^{\theta}. \qquad (3.62)$$

From the original Eq. (3.56) we have, after multiplying by $2x^2$,

$$2x^2 \frac{d^2\theta}{dx^2} + 2x \frac{d\theta}{dx} = -2Ax^2 e^{\theta}. \tag{3.63}$$

Equating (3.62) and (3.63) results in a simpler θ equation

$$\frac{d^2\theta}{dx^2} - \frac{1}{x}\frac{d\theta}{dx} - \frac{1}{2}\left(\frac{d\theta}{dx}\right)^2 = 0. \tag{3.64}$$

Equation (3.64) is easily integrated by the methods of Chapter 2. In this direction set $P = d\theta/dx$ so that Eq. (3.64) becomes the integrable Riccati equation

$$\frac{dP}{dx} - \frac{1}{x}P - \frac{1}{2}P^2 = 0 \tag{3.65}$$

with solution

$$P = \frac{d\theta}{dx} = -\frac{4Bx}{Bx^2 + 1} \tag{3.66}$$

where B is a constant of integration. A second integration gives

$$\theta(x) = C - 2\ln[Bx^2 + 1]. \tag{3.67}$$

There remains evaluation of the constants of integration, B and C.

To evaluate the constants B and C we turn to the boundary conditions and find that the condition $d\theta/dx = 0$ at $x = 0$ has already been used. Applying $\theta = 0$ at $x = 1$ allows us to determine

$$C = 2\ln(B + 1)$$

so that

$$\theta(x) = 2\ln[(B + 1)/(Bz^2 + 1)]. \tag{3.68}$$

The evaluation of B is carried out by requiring that Eq. (3.68) satisfy the original equation (3.56). Performing this substitution we find that $8B = A(B + 1)^2$ which has two roots given by

$$\{8 - 2A \pm [(8 - 2A)^2 - 4A^2]^{1/2}\}/2A.$$

Apparently this method of Chambré cannot be extended to the case of spherical symmetry ($n = 2$). As we shall see, in the next section, that equation is a special case of the Lane–Emden equation. Only one particular solution for $n = 2$ is known, namely $\theta(x) = \ln 2 - 2\ln x$.

An alternate treatment of Eq. (3.56) was carried out by Lempke [22].

3.4 Equations of Lane-Emden Type

By an equation of Lane–Emden type we shall mean any equation of the form

$$\frac{d^2 y}{dx^2} + \frac{2}{x}\frac{dy}{dx} + f(y) = 0 \tag{3.69}$$

where $f(y)$ is some given function of y. This equation was probably first studied by Emden [23] in examining the thermal behavior of spherical clouds of gas acting in gravitational equilibrium and subject to the laws of thermodynamics. A more recent treatment of the case with $f(y) = y^n$ is available in Chandrasekhar [24]. We sketch a brief derivation for $f = y^n$.

Consider a spherical cloud of gas and denote its total pressure at a distance r from the center by $p = p(r)$. The total pressure is due to the usual gas pressure plus a contribution from radiation

$$p = \tfrac{1}{3} a T^4 + RT/v \tag{3.70}$$

where a, T, R, and $v(r)$ are respectively radiation constant, absolute temperature, gas constant, and volume. Pressure and density $\rho = v^{-1}$ both vary with r and the relation between the first two is

$$p = K\rho^\gamma \tag{3.71}$$

where γ and K are constants. Let $m(r)$ be the mass within a sphere of radius r and G be the constant of gravitation. The equilibrium equations for the configuration are

$$\frac{dp}{dr} = -\frac{Gm\rho}{r^2} \tag{3.72a}$$

and†

$$\frac{dm}{dr} = 4\pi r^2 \rho. \tag{3.72b}$$

---

† An alternate derivation is available in the work of Davis [25].

To consolidate Eqs. (3.72a) and (3.72b) into one equation we eliminate m between the two and obtain

$$\frac{1}{r^2}\frac{d}{dr}\left(\frac{r^2}{\rho}\frac{dp}{dr}\right) + 4\pi G\rho = 0. \tag{3.73}$$

Now write $\gamma = 1 + \mu^{-1}$ and set

$$\rho = \lambda\phi^\mu \tag{3.74}$$

whereupon

$$p = K\rho^{1+1/\mu} = K\lambda^{(1+1/\mu)}\phi^{\mu+1}. \tag{3.75}$$

The constant λ is, for the moment, undetermined. When Eqs. (3.74) and (3.75) are substituted into Eq. (3.73) we find

$$\frac{1}{r^2}\frac{d}{dr}\left(r^2\frac{d\phi}{dr}\right) + k^2\phi^\mu = 0 \tag{3.76}$$

with

$$k^2 = 4\pi G\lambda^{(1-1/\mu)}/(\mu+1)K.$$

The final substitution of $x = kr$ transforms Eq. (3.76) into

$$\frac{d^2\phi}{dx^2} + \frac{2}{x}\frac{d\phi}{dx} + \phi^\mu = 0 \tag{3.77}$$

an equation of the form (3.69). If $\mu = 0$ (an impossible physical case) or 1 the equation is linear, otherwise it is nonlinear.

Hitherto λ was arbitrary. If we let $\lambda = \rho_0$, the density at $r = 0$, then we may take $\phi = 1$ at $x = 0$ as one auxiliary condition. By symmetry the other condition is $d\phi/dx = 0$ when $x = 0$. A solution of Eq. (3.77) satisfying these auxiliary conditions is called a Lane–Emden function of index $\mu = (\gamma - 1)^{-1}$. Only one case, that of $\mu = 5$ is exactly integrable. In all others, numerical or approximate solutions are necessary and tables are available for the cases $\mu = 0$ to $\mu = 6$ (see Miller [26]).

Since $\mu = 5$ corresponds to a physically realizable case $\gamma = 1.2$ we sketch the manner of solution. Eq. (3.77) may be expressed as

$$\frac{1}{x^2}\frac{d}{dx}\left(x^2\frac{d\phi}{dx}\right) + \phi^\mu = 0, \tag{3.78}$$

which form suggests to us the operator transformation†

$$\frac{d}{d\eta} = -x^2 \frac{d}{dx},$$

that is to say $x = \eta^{-1}$. Under this transformation Eq. (3.78) becomes the equivalent form

$$\eta^4 \frac{d^2\phi}{d\eta^2} + \phi^\mu = 0. \tag{3.79}$$

In its turn Eq. (3.79) can be transformed to a more amenable form by setting

$$\phi = A\eta^n y(\eta) \tag{3.80}$$

in which A and n are adjustable parameters. Thus Eq. (3.79) becomes

$$\eta^{4+n} y'' + 2n\eta^{3+n} y' + n(n-1)\eta^{n+2} y + A^{\mu-1}\eta^{\mu n} y^\mu = 0.$$

When this equation is divided by η^{n+2} and n is chosen as $\frac{1}{2}$ there results

$$\eta^2 y'' + \eta y' - \tfrac{1}{4} y[1 - 4A^{\mu-1}\eta^{(\mu-5)/2} y^{\mu-1}] = 0. \tag{3.81}$$

The last term becomes independent of the independent variable η when $\mu = 5$ and is further simplified by taking $A = (\tfrac{1}{4})^{1/4}$. Further, the form of the two leading terms suggests an Euler equation. Therefore we set $\eta = e^u$ and thus Eq. (3.81) now becomes

$$\frac{d^2 y}{du^2} - \frac{1}{4} y(1 - y^4) = 0 \tag{3.82}$$

which is free of the independent variable u.

To integrate Eq. (3.82) we use the transformation $v = dy/du$, $d^2y/du^2 = v \, dv/dy$ whereupon Eq. (3.82) becomes

$$v \frac{dv}{dy} = \frac{1}{4} y(1 - y^4).$$

The first integral is

$$v^2 = \tfrac{1}{4} y^2 (1 - \tfrac{1}{3} y^4) + C. \tag{3.83}$$

---

† The $-$ sign is helpful in what follows.

Now $r = 0$ implies $x = 0$ at which point $d\phi/dx = 0$. This implies that at $y = 0$, $v = 0$ so that $C = 0$. Thus Eq. (3.83) becomes

$$\frac{dy}{du} = \pm \frac{1}{2} y \left[1 - \frac{1}{3} y^4 \right]^{1/2}. \tag{3.84}$$

As a matter of fact either sign will lead to the final result but we choose the negative sign so that when $u \to \infty$, $\eta \to \infty$ or $x \to 0$.

The integration of Eq. (3.84) is completed by setting

$$\frac{1}{3} y^4 = \sin^2 \omega = 4 \tan^2 \frac{\omega}{2} \Big/ \left(1 + \tan^2 \frac{\omega}{2} \right)^2. \tag{3.85}$$

Thus

$$u = -2 \int \frac{dy}{y} \frac{1}{(1 - \frac{1}{3} y^4)^{1/2}} + B = - \int \frac{d\omega}{\sin \omega} + B$$

$$= -\ln[\tan(\omega/2)] + B.$$

This result is expressible in the alternate form

$$\tan(\omega/2) = De^{-u} = D\eta^{-1} = Dx. \tag{3.86}$$

When Eq. (3.86) is substituted into Eq. (3.85) and fourth roots taken we find

$$y = [12D^2 x^2/(1 + D^2 x^2)^2]^{1/4}. \tag{3.87}$$

But

$$\phi = (\tfrac{1}{4})^{1/4} \eta^{1/2} y, \qquad x = \eta^{-1} \qquad \text{so that} \quad \phi = (\tfrac{1}{4})^{1/4} x^{-1/2} y.$$

Upon setting these into Eq. (3.87) we find

$$\phi = [3D^2/(1 + D^2 x^2)^2]^{1/4}. \tag{3.88}$$

To evaluate D we apply the boundary condition $\phi = 1$ when $x = 0$ and obtain $D^2 = \frac{1}{3}$ whereupon the Lane–Emden function of index 5 has the form

$$\phi_5 = (1 + \tfrac{1}{3} x^2)^{-1/2}.$$

The remaining functions are obtained by series expansions since apparently they are not expressible in terms of elementary functions.

3.5 Some Similarity Equations from Fluid Mechanics

A number of conspicuous advances in fluid mechanics have occurred because of similarity solutions. We shall not herein discuss the general theory. Theory and examples, in various disciplines, are given by Hansen [27], Ames [28], Kline [29], and Sedov [30]. The ordinary differential equations which arise from the construction of similarity variables are often of intractable form. The difficulties are compounded because the boundary conditions occur in part at infinity.†

This section begins with several examples whose exact solutions are attainable. It concludes with a discussion of the conversion of equations to integral equation form.

A. PLANE HYDRODYNAMIC JET

Bickley [31] and Schlichting [32] consider the problem of the plane hydrodynamic jet. In that theory occurs the equation

$$3\varepsilon \frac{d^3 y}{dx^3} + y \frac{d^2 y}{dx^2} + \left(\frac{dy}{dx}\right)^2 = 0, \qquad (3.89)$$

subject to the boundary conditions $y = y'' = 0$ when $x = 0$ and $y' = 0$ when $x \to \infty$. This equation is solvable not only with these boundary conditions but in more general cases.

Equation (3.89) is expressible in the form

$$\frac{d}{dx}[3\varepsilon y'' + yy'] = 0$$

whence a first integration yields

$$3\varepsilon y'' + yy' = A. \qquad (3.90)$$

As a result of the boundary conditions at $x = 0$ we see that $A = 0$ so that

$$3\varepsilon y'' + yy' = 0$$

---

† In Section 3.6 we discuss the conversion of boundary value problems into initial value problems. In the Appendix the development of similarity variables by means of groups is sketched.

or

$$\frac{d}{dx}\left[3\varepsilon y' + \frac{1}{2}y^2\right] = 0$$

which integrates to

$$6\varepsilon y' + y^2 = B. \tag{3.91}$$

For convenience we set $B = a^2$ and find that Eq. (3.91) can be rewritten as

$$6\varepsilon \int \frac{dy}{a^2 - y^2} = x + C$$

which integrates to

$$\frac{6\varepsilon}{a} \tanh^{-1}(y/a) = x + C.$$

Now $y = 0$ when $x = 0$ so that $C = 0$, whereupon

$$y = a \tanh\left(\frac{ax}{6\varepsilon}\right) \tag{3.92}$$

which satisfies the three boundary conditions. Equation (3.92) is a function of the constant $B = a^2$.

It should also be noted that Eq. (3.90) can be integrated when A, B, and C are all not zero. For commencing with Eq. (3.90) we see that it is expressible as

$$\frac{d}{dx}\left[3\varepsilon y' + \frac{1}{2}y^2\right] = A$$

so that

$$3\varepsilon y' + \tfrac{1}{2}y^2 = Ax + B. \tag{3.93}$$

Let $u = Ax + B$ whereupon

$$\frac{dy}{dx} = \frac{dy}{du}\frac{du}{dx} = A\frac{dy}{du}$$

so Eq. (3.93) becomes

$$\frac{dy}{du} + \frac{1}{6\varepsilon A}y^2 - \frac{u}{3\varepsilon A} = 0$$

which is a Riccati equation of integrable form with the solution as described by Eqs. (2.69) and (2.70).

B. BOUNDARY LAYER FLOW IN A CONVERGENT CHANNEL

Falkner and Skan [33] found that similar solutions exist for the boundary layer equations

$$uu_x + vu_y = U\frac{dU}{dx} + vu_{yy}$$

$$u_x + v_y = 0$$

(3.94)

with $u = v = 0$ for $y = 0$ and $u = U(x) = cx^m$ for $y \to \infty$, if the stream function $f(\eta)$, $\eta = y(Re)^{1/2}/Lx^{1/2}$, satisfies the ordinary differential equation

$$f''' + \alpha ff'' + \beta(1 - f'^2) = 0.$$

(3.95)

The boundary conditions on f are

$$f = f' = 0 \quad \text{at} \quad \eta = 0, \qquad f' = 1 \quad \text{at} \quad \eta \to \infty.$$

Equation (3.95) is usually not solvable exactly. However for the case of flow in a convergent channel $\alpha = 0$ and $\beta = +1$ whereupon f satisfies

$$f''' + 1 - f'^2 = 0$$

(3.96)

and this is integrable. Admittedly, this is one of the rare cases when the solution of the boundary layer equations can be obtained in closed form.

We describe the solution of Eq. (3.96) as obtained by Pohlhausen [34] (see also Schlichting [32]). Upon multiplying Eq. (3.96) by f'' and integrating we have

$$(f'')^2 - \tfrac{2}{3}(f' - 1)^2(f' + 2) = A$$

(3.97)

where A is a constant of integration. The value of A is zero since $f' = 1$ for $\eta \to \infty$ and therefore $f'' \to 0$ as $\eta \to \infty$. Thus

$$\eta = \left(\frac{3}{2}\right)^{1/2} \int_0^{f'} \frac{df'}{\{(f' - 1)^2(f' + 2)\}^{1/2}}.$$

(3.98)

If only the velocity components u and v are desired then only the integration of Eq. (3.98) is required since

$$f' = u/U, \qquad v = -(vu_1)^{1/2} \eta f'/x$$

where

$$\eta = (y/x)(u_1/v)^{1/2}, \qquad U = u_1/(-x).$$

The integral of Eq. (3.98) is expressible in closed form as

$$\eta = \sqrt{2}\left\{\tanh^{-1}\frac{\sqrt{2}+f'}{\sqrt{3}} - \tanh^{-1}\sqrt{\frac{2}{3}}\right\}$$

or solving for f'

$$f' = u/U = 3\tanh^2\left[\frac{\eta}{\sqrt{2}} + 1.146\right] - 2. \qquad (3.99)$$

When the stream function $\psi = -\sqrt{vu_1}\,f(\eta)$ is required another integration of Eq. (3.99) must be performed. For small α, Eq. (3.99) may be used as the zero order term in a perturbation solution (see Chapter 4).

C. Laminar Circular Jet

We indicate Schlichting's [35] solution for the laminar circular jet which is the generalization of the problem of part a of this section. The boundary equations in this case are

$$uu_x + vu_y = vy^{-1}\frac{\partial}{\partial y}(y^{-1}u_y)$$

$$u_x + v_y + vy^{-1} = 0 \qquad\qquad (3.100)$$

with boundary conditions

$$v = 0, \quad \frac{\partial u}{\partial y} = 0 \quad \text{at} \quad y = 0, \quad u = 0 \quad \text{at} \quad y \to \infty.$$

One obtains the similarity variables as

$$\eta = y/x, \quad \psi = vx\,F(\eta)$$

where ψ is the stream function. From these results we obtain the velocity components

$$u = vF'/x\eta, \quad v = \frac{v}{x}\left[F' - \frac{F}{\eta}\right].$$

When these are inserted into Eq. (3.100) we find the stream function equation to be

$$\frac{FF'}{\eta^2} - \frac{F'^2}{\eta} - \frac{FF''}{\eta} = \frac{d}{d\eta}\left[F'' - \frac{F'}{\eta}\right]. \tag{3.101}$$

The left hand side is the η derivative of $-FF'/\eta$. Thus a first integral of Eq. (3.101) is

$$FF' = F' - \eta F'' + C. \tag{3.102}$$

The boundary conditions are $u = u_m$ and $v = 0$ for $y = 0$. Hence $F' = 0$, $F = 0$ for $\eta = 0$ so $C = 0$.

A second constant of integration can be evaluated as follows: If $F(\eta)$ is a solution of Eq. (3.102) (with $C = 0$), then $F(\gamma\eta) = F(\phi)$ is also a solution as an elementary verification demonstrates. A particular solution of the equation

$$F\frac{dF}{d\phi} = \frac{dF}{d\phi} - \phi\frac{d^2F}{d\phi^2} \tag{3.103}$$

subject to the boundary conditions $F = F' = 0$ at $\phi = 0$ is now required. The key to this integration is to write Eq. (3.103) as

$$\phi^2\frac{d^2F}{d\phi^2} + (F - 1)\phi\frac{dF}{d\phi} = 0 \tag{3.104}$$

obtained by multiplying by ϕ. Again this suggests the Euler transformation $\phi = e^u$ whereupon

$$\phi\frac{dF}{d\phi} = \frac{dF}{du}, \qquad \phi^2\frac{d^2F}{d\phi^2} = \frac{d^2F}{du^2} - \frac{dF}{du}.$$

Performing this transformation we find that Eq. (3.104) becomes

$$\frac{d^2F}{du^2} + (F - 2)\frac{dF}{du} = 0. \tag{3.105}$$

With $p = dF/du$, $d^2F/du^2 = p\,dp/dF$ this equation reduces to

$$\frac{dp}{dF} = 2 - F$$

or $p = 0$. The latter is discarded. The solution is given by

$$F = \frac{\phi^2}{1 + \phi^2/4}. \tag{3.106}$$

To evaluate γ we cannot use standard methods but apply the assumption of constant flux of momentum in the x direction

$$J = \text{const.} = 2\pi\rho \int_0^\infty u^2 y \, dy.$$

From the defining relation for u we find

$$u = \frac{\nu}{x}\gamma^2 \frac{1}{\phi}\frac{dF}{d\phi} = \frac{\nu}{x}\frac{2\gamma^2}{(1 + \phi^2/4)^2}$$

where $\phi = \gamma y/x$. Thus $J = (16/3)\,\pi\rho\gamma^2\nu^2$.

D. Similarity in Gas Dynamics

The unsteady, one dimensional, anisentropic flow of a polytropic gas, neglecting the effects of viscosity, conduction and radiation is represented by the Euler equations

$$\rho u_t + \rho u u_x + p_x = 0 \tag{3.107a}$$

$$\rho_t + (\rho u)_x = 0 \tag{3.107b}$$

$$S_t + u S_x = 0 \tag{3.107c}$$

$$S = C_v \ln[p\rho^{-\gamma}] \tag{3.107d}$$

where ρ, p, u, and S are the density, pressure, velocity, and entropy respectively. C_v is the specific heat at constant volume, γ is the ratio of specific heats and $\tau = 1/\rho$ is the specific volume.

We first illustrate the introduction of an auxiliary function ϕ which permits the consolidation of the system (3.107) into one equation. The simple definition $\phi_x = \rho$, $\phi_t = -\rho u$, suggested by Eq. (3.107b), does not apply since it does not provide for sufficient degrees of freedom. The required freedom is available only at the *second derivative level*. Upon multiplying Eq. (3.107b) by u and adding to Eq. (3.107a) we have the equation

$$(\rho u)_t + (p + \rho u^2)_x = 0 \tag{3.108}$$

as a replacement for Eq. (3.107a). To assure satisfaction of Eq. (3.107b) (continuity) set

$$\phi_{xx} = \rho, \qquad \phi_{xt} = -\rho u \tag{3.109}$$

whereupon Eq. (3.108) becomes $-\phi_{xtt} + (p + \rho u^2)_x = 0$. This suggests defining

$$\phi_{tt} = p + \rho u^2, \qquad (3.110)$$

that is, Eq. (3.108) is now automatically satisfied. We now have the additional definitions

$$u = -\phi_{xt}/\phi_{xx} \qquad (3.111)$$

$$p = \phi_{tt} - \rho u^2 = (\phi_{tt}\phi_{xx} - \phi_{xt}^2)/\phi_{xx} \qquad (3.112)$$

$$S = C_v \ln\left[\frac{\phi_{tt}\phi_{xx} - \phi_{xt}^2}{\phi_{xx}^{1+\gamma}}\right]. \qquad (3.113)$$

There remains the energy Eq. (3.107c) which is expressible, because of Eq. (3.111) as

$$\phi_{xx}S_t - \phi_{xt}S_x = 0. \qquad (3.114)$$

This equation is identically satisfied if we take

$$S = f(\phi_x) \qquad (3.115)$$

where f is an arbitrary function, having the necessary differentiability properties. Returning to Eq. (3.113) and taking exponentials the equation for ϕ becomes

$$\phi_{xx}\phi_{tt} - \phi_{xt}^2 = F(\phi_x)\,\phi_{xx}^{1+\gamma} \qquad (3.116)$$

where $F(\phi_x) = \exp[f(\phi_x)/C_v]$. If the flow is isentropic then $F(\phi_x)$ is a constant. In what follows we shall use the more general relation

$$F(\phi_x) = k\phi_x^n. \qquad (3.117)$$

Studies of Eq. (3.116) have occupied numerous researchers (see Ames [28]). Herein our concern is with the possible similarity solutions. Ames [36] has applied the group theory procedure to ascertain the similarity possibilities. The resulting equations are

$$\rho(x, t) = t^{A-2B}f''(\eta)$$
$$u(x, t) = t^{B-1}[B\eta - (A - B)f'/f'']$$
$$p(x, t) = \frac{t^{A-2}}{f''}\{A(A - 1)ff'' + (1 - 2B)B\eta f'f'' - (A - B)^2 f'^2\} \qquad (3.118)$$
$$S(x, t) = k[t^{A-B}f'(\eta)]^n$$

where

$$\eta = x/t^B, \qquad \phi(x, t) = t^A f(\eta)$$

and f satisfies the differential equation

$$[A(A - 1)f + B\eta(1 - B)f']f'' - (A - B)^2 f'^2 = k(f')^n (f'')^{1+\gamma}. \qquad (3.119)$$

One, but not both, of the parameters A and B is available for our arbitrary selection. A and B are related by the group invariance requirement

$$B = \frac{A(n + \gamma - 1) + 2}{n + 2\gamma}. \qquad (3.120)$$

A variety of simplifications seem possible because of the availability of one parameter. The choice of $A = 1$ eliminates one term but requires $B = (n + \gamma + 1)/(n + 2\gamma)$ and of $B = 1$ eliminates one term but requires $A = (n + 2\gamma - 2)/(n + \gamma - 1)$. If we set $A = B$ we find their common value must be $2/(\gamma + 1)$. Lastly, $A = B = 1$ is possible for $\gamma = 1$. Some of the resulting equations lead to trivial results and some are very challenging. Most cases require numerical or approximate solution techniques. We concentrate our attention on isentropic flow with $n = 0$ and without loss of generality let $k \equiv 1$ in Eq. (3.119).

A simplification of Eqs. (3.118) occurs if we set $A = 2$ and $B = 1$ but of course this may not be possible because of Eq. (3.120), to say nothing of physical reality. However, with $n = 0$, $A = 2$ we find that $B = 1$ is satisfied for all γ. In this case the similarity variables become

$$\eta = x/t \qquad \text{and} \qquad \phi = t^2 f(\eta)$$

where f satisfies the differential equation

$$2ff'' - (f')^2 = (f'')^{1+\gamma}. \qquad (3.121)$$

A variety of equations whose left hand terms are the same are tabulated by Kamke [37 Eqs. (6.148)–(6.154)]. These equations are often *solvable by differentiation* as we shall see in the sequel.

Upon taking the η derivative of Eq. (3.121) we find

$$2ff''' + 2f'f'' - 2f'f'' = (1 + \gamma)(f'')^\gamma f'''$$

which can be rearranged to

$$f'''[2f - (1 + \gamma)(f'')^\gamma] = 0. \qquad (3.122)$$

Equation (3.122) is equivalent to the two equations

$$f''' = 0$$
$$f'' = [2f/(1 + \gamma)]^{1/\gamma}. \qquad (3.123)$$

The first of these is immediately integrable to

$$f = a\eta^2 + b\eta + c$$

a form which leads to the essentially trivial *physical* solution, calculated from Eqs. (3.118), $\rho = $ const. $= 2a$, $u = $ const., and $p = $ const.

The second form of Eq. (3.123), is more interesting. We write it in the more easily handled form

$$f'' = Af^p$$

where $A = (2/(1 + \gamma))^{1/\gamma}$ and $p = 1/\gamma$. In the most useful physical problems $\gamma \geq 1$ so that usually $p \leq 1$. Setting $v = f'$, $f'' = v\, dv/df$ whereupon we have

$$v \frac{dv}{df} = Af^p$$

whose integral is

$$v^2 = \frac{2A}{p + 1} f^{p+1} + C. \qquad (3.124)$$

If $f'(0) = f(0) = 0$ we find $C = 0$. Therefore Eq. (3.124) becomes

$$\frac{df}{d\eta} = \pm \left(\frac{2A}{p + 1} \right)^{1/2} f^{(p+1)/2} \qquad (3.125)$$

which integrates to

$$f(\eta) = \left[\frac{1}{2} \left(\frac{\gamma - 1}{\gamma} \right) \left(\frac{2\gamma}{\gamma + 1} \right)^{1/2} \left(\frac{2}{\gamma + 1} \right)^{1/2\gamma} \eta + D \right]^{(\gamma - 1)/2\gamma}. \qquad (3.126)$$

The evaluation of D now proceeds via the condition $f(0) = 0$.

The case of $\gamma = p = 1$ deserves mention. Here Eq. (3.123) becomes $f'' - f = 0$ whose solution is

$$f(\eta) = C \exp(x/t) + D \exp(- x/t).$$

E. Conversion to Integral Equations

There are several excellent advantages that accrue to the analyst when a differential equation is converted to an integral equation. One of these is the well-known "smoothing" property of the integral form. A second results from the ease of inclusion of boundary conditions. A third advantage lies in the relative ease of development of asymptotic solutions as discussed in Chapter 4.

We begin with an example due to Lock [38]. From Eq. (3.95) with $\alpha = 1$, $\beta = 0$ we obtain the famous Blasius equation,

$$f''' + ff'' = 0 \tag{3.127}$$

where $f(0) = f'(0) = f'(\infty) - 2 = 0$, for determining a similarity solution to the problem of steady laminar boundary layer flow over a semiinfinite flat plate. Upon rewriting Eq. (3.127) as

$$df''/f'' = -f \, d\eta$$

and integrating from 0 to η, we obtain†

$$f'' = \alpha \exp\left[-\int_0^\eta f(u) \, du \right] \tag{3.128}$$

where the constant of integration α determines the shear stress on the plate. Integrating twice more we find

$$f = \alpha \int_0^\eta \int_0^\eta \exp\left[-\int_0^\eta f \, d\eta \right] d\eta \, d\eta + \beta_1 + \beta_2 \eta.$$

Satisfaction of the two conditions $f(0) = f'(0) = 0$ at the plate requires $\beta_1 = \beta_2 = 0$. Thus the integral form of Eq. (3.127), satisfying two boundary conditions is

$$f = \alpha \int_0^\eta \int_0^\eta \left[\exp -\int_0^\eta f \, d\eta \right] d\eta \, d\eta$$

$$= \alpha \int_0^\eta (\eta - u) \left[\exp -\int_0^u f \, du \right] du. \tag{3.129}$$

To achieve the final result of Eq. (3.129) we have utilized the easily proved

---

† u is a dummy variable of integration.

result (see, e.g., Hildebrand [39])

$$\int_a^x \underbrace{\cdots \int_a^x f(x)\, \underbrace{dx \cdots dx}_{n \text{ times}}}_{n \text{ times}} = \frac{1}{(n-1)!} \int_a^x (x-y)^{n-1} f(y)\, dy.$$

The method of solution described herein is not generally applicable unless the governing equation possesses the property that *a pair of dependent variables involving the function and its derivatives* can be *expressed in the form*

$$f^{(p)} = F[f^{(q)}, \eta], \qquad q < p. \tag{3.130}$$

The "solution" for $f^{(q)}$, given by

$$f^{(q)}(\eta) = \frac{1}{(p-q-1)!} \int_0^\eta (\eta - u)^{p-q-1} F[f^{(q)}, u]\, du$$

$$+ \sum_{k=1}^{p-q} \beta_k \eta^{k-1}/(k-1)! \tag{3.131}$$

can then be found by successive approximations satisfying some of the boundary conditions through $\beta_1, \beta_2, \ldots, \beta_{p-q}$. The final determination of $f(\eta)$ proceeds in a related manner. More will be said about this problem in Chapter 4.

Certainly all similarity equations, even in boundary layer theory, are not amenable to such a procedure. Alternate but related methods of conversion are available and can be used. In this connection we mention the fundamental work of Weyl [40] who discusses the solvability of the nonlinear boundary value problem

$$w'''(z) + 2w(z)w''(z) + 2\lambda\{k^2 - w'^2(z)\} = 0, \qquad z > 0$$
$$w(0) = w'(0) = 0$$
$$w'(z) \to k, \qquad z \to \infty \tag{3.132}$$
$$k = \text{const}, \quad \lambda = \text{const} \geq 0.$$

In the following material we shall concern ourselves with the problem of Homann [41] where $\lambda = \frac{1}{2}$, $k = 1$, which corresponds to steady flow in the boundary layer along a surface of revolution near the forward stagnation point. Homann considered the problem by series and asymptotic methods. The conversion to an integral equation problem was done by Siekmann [42].

The present problem concerns the equation

$$w''' + 2ww'' + \{1 - w'^2(z)\} = 0 \tag{3.133a}$$

$$w(0) = w'(0) = 0, \qquad w'(\infty) = 1. \tag{3.133b}$$

Upon examination of Eq. (3.133a) we note that it contains the two terms $2ww''$ and $-w'^2$ which, in part, counteract one another upon differentiation. Thus, by differentiation we obtain†

$$w'''' + 2ww''' = 0 \tag{3.134}$$

which is exactly of the general form Eq. (3.130).

Let us now denote by $w = f(z)$ that solution of Eq. (3.134) which has the *initial values*

$$0 = f(0) = f'(0), \qquad f''(0) = 1, \qquad f'''(0) = -\beta^2 \tag{3.135}$$

and also satisfies the third order equation

$$f''' + 2ff'' + (\beta^2 - f'^2) = 0. \tag{3.136}$$

Upon dividing Eq. (3.134) by w''' and integrating from 0 to z we have

$$f'''(z) = -\beta^2 \exp\left[-2\int_0^z f(t)\, dt\right]. \tag{3.137}$$

Our eventual goal is an integral equation for f''. On the way to this goal we note that for *our f*, with the boundary conditions (3.135),

$$\int_0^z f(t)\, dt = \tfrac{1}{2}\int_0^z (z - t)^2 f''(t)\, dt, \tag{3.138}$$

a result which is easily verified by parts integrations. Using Eq. (3.138) we find that Eq. (3.137) becomes

$$f'''(z) = -\beta^2 \exp\left[-\int_0^z (z - t)^2 f''(t)\, dt\right]. \tag{3.139}$$

If a function $g(z)$ is defined by setting

$$f''(z) = g(z)$$

---

† Note that Eq. (3.134), as well as the Blasius equation, is invariant under the transformation $w(z) \to \varepsilon w(\varepsilon z)$ where ε is constant. This property allows a reduction of order. ε is called a constant of "homology."

an integration of Eq. (3.139) leaves

$$g(z) = 1 - \beta^2 \int_0^z \exp\left[-\int_0^\sigma (\sigma - t)^2 g(t)\, dt \right] d\sigma. \qquad (3.140)$$

This is clearly a nonlinear integral equation for $g(z)$.

The constant β can be calculated by using the condition $g(\infty) = 0$ which results from the original asymptotic boundary condition $w'(\infty) = 1$. From this we obtain

$$\beta^2 = \int_0^\infty \exp\left[-\int_0^\sigma (\sigma - t)^2 g(t)\, dt \right] d\sigma. \qquad (3.141)$$

With this value of β we have completed the introduction of all boundary conditions. The solution of the integral Eq. (3.140) is then used to determine

$$\begin{aligned}
f(z) &= \int_0^z f'(t)\, dt \\
&= \int_0^z (z - t) f''(t)\, dt \\
&= \int_0^z (z - t) g(t)\, dt. \qquad (3.142)
\end{aligned}$$

Having determined f the solution of the original problem described by Eq. (3.133) is given by

$$w(\eta) = \varepsilon f(z), \qquad \eta = z/\varepsilon$$

where ε is selected so that $w' \to 1$ as $z \to \infty$. Thus $\varepsilon = \beta^{-1/2}$.

A further application of these concepts has been carried out by Siekmann [43] for a problem of the laminar boundary layer along a flat plate in which a portion of the plate between $0 \le x < L_0$ is at rest and the surface of the plate between $L_0 \le x \le L$ is moving with a constant velocity.

3.6 Conversion of Boundary Value to Initial Value Problems

The conversion of partial differential equations to ordinary differential equations by means of similarity transformations usually results in a boundary value problem with impressed conditions at some finite value (usually zero) and at ∞. Such boundary value problems† are difficult. On the other hand, initial value problems, with all auxiliary conditions specified at one point, are easier to analyze and are especially suited for approximate and numerical solution. All we must do for initial value problems is "march" the solution out from specified initial conditions.

The reader has probably already noticed that the success of the Weyl–Siekmann method (Section 3.5e, Eqs. (3.133)–(3.140)) depended strongly upon conversion to the initial value problem specified by Eqs. (3.135) and (3.136). In this section we discuss the conversion of certain boundary value problems into initial value problems. The origin of this concept lies in work of Blasius (see Schlichting [32]) and the general form is due to Klamkin [44]. We introduce the procedure with the Blasius problem

$$f''' + ff'' = 0, \qquad f = f(\eta) \tag{3.143a}$$

$$f(0) = f'(0) = 0, \qquad f(\infty) = 2 \tag{3.143b}$$

before describing the generalization of Klamkin.

Let us introduce an unspecified parameter λ and replace $f'(\infty) = 2$ with the initial condition $f''(0) = \lambda$. With the values $f(0) = f'(0) = 0$, $f''(0) = \lambda$ we develop a Taylor series‡ solution about $\eta = 0$ for Eq. (3.143a) as

$$f(\eta) = \sum_{k=0}^{\infty} f^{(k)}(0)\eta^k/k!. \tag{3.144}$$

From the initial conditions and Eq. (3.143a) we have

$$f(0) = 0, \qquad f'(0) = 0, \qquad f''(0) = \lambda,$$

$$f'''(0) = -f(0)f''(0) = 0,$$

$$f^{(4)} = -(ff'')' = -f'f'' - ff''',$$

$$f^{(4)}(0) = 0, \quad \text{etc.}$$

The series is

$$f(\eta) = \frac{\lambda\eta^2}{2!} - \frac{\lambda^2\eta^5}{5!} + \frac{11\lambda^3\eta^8}{8!} - \frac{375\lambda^4\eta^{11}}{11!} \cdots. \tag{3.145}$$

---

† Both analytically and numerically initial value problems are more amenable than boundary value problems.

‡ Further amplification of series solutions will be given in Chapter 4.

Let $F(\eta)$ represent the solution corresponding to $\lambda = 1$ in Eq. (3.145), that is

$$F(\eta) = \frac{\eta^2}{2!} - \frac{\eta^5}{5!} + \frac{11\eta^8}{8!} - \frac{375\eta^{11}}{11!} \cdots. \qquad (3.146)$$

Consequently, we see that

$$f(\eta) = \lambda^{1/3} F(\lambda^{1/3}\eta) \qquad (3.147)$$

and

$$2 = \lim_{\eta \to \infty} f'(\eta) = \lambda^{2/3} \lim_{\eta \to \infty} F'(\lambda^{1/3}\eta) = \lambda^{2/3} \lim_{\eta \to \infty} F'(\eta). \qquad (3.148)$$

Thus the value of the initially undetermined parameter λ is

$$\lambda = \{2/F'(\infty)\}^{3/2}.$$

Let us summarize what has been accomplished. The original boundary value problem Eqs. (3.143a) and (3.143b) has been transformed into a *pair* of initial value problems given by

$$F''' + FF'' = 0$$
$$F(0) = F'(0) = 0, F''(0) = 1 \qquad (3.149)$$

and

$$f''' + ff'' = 0$$
$$f(0) = f'(0) = 0, \quad f''(0) = \{2/F'(\infty)\}^{3/2}. \qquad (3.150)$$

The reduction is advantageous for numerical computation since it avoids any interpolation techniques. We also remark that the expansion into series, Eq. (3.145), is not necessary but was done for motivation.

This idea of Blasius is generalizable to a wide class of differential equations of any order where the boundary conditions are specified at $\eta = 0$ and at $n = \infty \, (-\infty)$. We first consider the (rather) general second order equation, then sketch the results for a general third order equation. Last we examine the possibilities for a *system* of second order equations.

Suppose the rather general second order equation

$$\sum_{m,n,r,s} A_{mnrs} (f'')^m (f')^n f^r \eta^s = 0 \qquad (3.151)$$

is subject to the boundary conditions $f(0) = 0$, $f^{(d+1)}(\infty) = k$. Here m, n, r, and s are arbitrary indices, d is an arbitrary integer (0 or ± 1 here), and A_{mnrs} are arbitrary constants.

Let $f'(0) = \lambda$ and assume that f can be expressed in the form

$$f(\eta) = \lambda^{1-\alpha} F(\lambda^{\alpha} \eta) \tag{3.152}$$

where $F(\eta)$ satisfies Eq. (3.151) but subject to the initial conditions $F(0) = 0$, $F'(0) = 1$. In order that both f and F satisfy Eq. (3.151) there must exist some dimensional homogeneity on the exponents m, n, r, and s. To obtain this condition we substitute Eq. (3.152) into Eq. (3.151) and obtain

$$\sum_{m,n,r,s} A_{mnrs} [F''(\lambda^{\alpha}\eta)]^m [F'(\lambda^{\alpha}\eta)]^n [F(\lambda^{\alpha}\eta)]^r [\lambda^{\alpha}\eta]^s \lambda^c = 0 \tag{3.153}$$

where

$$c = (1 + \alpha)m + n + (1 - \alpha)r - \alpha s. \tag{3.154}$$

For Eq. (3.153) to be exactly Eq. (3.151), λ^c must factor out of *all terms*. That is c must be a constant for all m, n, r, and s. Consequently, the form of the allowable equation is not Eq. (3.151) but

$$\sum_{m,r,s} A_{mrs} [f'']^m [f']^n f^r \eta^s = 0 \tag{3.155}$$

where n is given by Eq. (3.154). Then

$$f^{(d+1)}(\infty) = k = \lambda^{1+d\alpha} F^{(d+1)}(\infty)$$

or

$$\lambda = \{k/F^{(d+1)}(\infty)\}^{1/(1+d\alpha)}. \tag{3.156}$$

In an entirely analogous fashion we can solve the third (and higher) order problem

$$\sum_{m,r,s,t} A_{mrst} [f''']^m [f'']^n [f']^s f^s \eta^t = 0$$
$$f(0) = f'(0) = 0, \qquad f^{(d+1)}(\infty) = k \tag{3.157}$$

where m, r, s, t are arbitrary indices, d is an arbitrary integer ($d = 0, \pm 1, 2$ here), and A_{mrst} are arbitrary constants. The quantity n, because of the homogeneity requirements, is given by

$$(1 + \alpha)m + n + (1 - \alpha)r + (1 - 2\alpha)s - \alpha t = c \quad \text{(const)}.$$

Here we have assumed that $f(\eta)$ can be expressed as

$$f(\eta) = \lambda^{1-2\alpha}F(\lambda^{\alpha}\eta)$$

where $f''(0) = \lambda$ and $F(\eta)$ satisfies Eq. (3.157) but with the initial conditions

$$F(0) = F'(0) = 0, F''(0) = 1.$$

Then

$$\lambda = \{k/F^{(d+1)}(\infty)\}^{1/(1+(d-1)\alpha)}.$$

These procedures are easily extended to equations of any order with similar type boundary conditions. In addition the same procedures apply to a wide class of simultaneous equations. As a motivation we examine a problem of Greenspan and Carrier [45] concerning the boundary layer flow of a viscous, electrically conducting incompressible fluid past a semiinfinite flat plate in the presence of a magnetic field. The boundary layer equations under similarity reduce to

$$\begin{aligned} f''' + ff'' - \beta gg'' &= 0 \\ g'' + \varepsilon(fg' - f'g) &= 0 \end{aligned} \tag{3.158}$$

with boundary conditions

$$f(0) = f'(0) = 0, \quad f'(\infty) = 2, \quad g(0) = 0, \quad g'(\infty) = 2. \tag{3.159}$$

The avoidance of interpolatory schemes is highly desirable. We therefore attempt to convert this boundary value problem into an initial value one by setting

$$\begin{aligned} f''(0) &= \lambda, & g'(0) &= \mu \\ f(\eta) &= \lambda^{1-2\alpha}F(\lambda^{\alpha}\eta), & g(\eta) &= \mu^{1-\gamma}G(\mu^{\gamma}\eta), \end{aligned} \tag{3.160}$$

where

$$F(0) = F'(0) = 0, \quad F''(0) = 1, \quad G(0) = 0, \quad G'(0) = 1, \text{ and } F(\eta),$$
$G(\eta)$ satisfy Eq. (3.158). Of course, this requires that the system

$$\lambda^{1+\alpha}F'''(\lambda^{\alpha}\eta) + \lambda^{2-2\alpha}F(\lambda^{\alpha}\eta)F''(\lambda^{\alpha}\eta) - \beta\mu^2 G(\mu^{\gamma}\eta)G''(\mu^{\gamma}\eta) = 0$$

$$\mu^{1+\gamma}G''(\mu^{\gamma}\eta) + \varepsilon\{\lambda^{1-2\alpha}\mu F(\lambda^{\alpha}\eta)G'(\mu^{\gamma}\eta) - \lambda^{1-\alpha}\mu^{1-\gamma}F'(\lambda^{\alpha}\eta)G(\mu^{\gamma}\eta)\} = 0$$

is the same as Eq. (3.158). For this to occur we must have

$$\left.\begin{array}{l} \lambda^{\alpha} = \mu^{\gamma} \\[6pt] \lambda^{1+\alpha} = \lambda^{2-2\alpha} = \mu^2 \\[6pt] \mu^{1+\gamma} = \lambda^{1-2\alpha} = \lambda^{1-\alpha}\mu^{1-\gamma}. \end{array}\right\} \tag{3.161}$$

Equations (3.161) are satisfied nontrivially by

$$\alpha = 1/3, \qquad \gamma = 1/2, \qquad \text{and} \qquad \mu = \lambda^{2/3}.$$

Clearly μ is *not independent* of λ so we find it impossible, in general, to satisfy both conditions at infinity. But if the initial conditions to Eq. (3.158) are

$$f(0) = f'(0) = 0, \qquad f''(0) = \lambda, \qquad g(0) = 0, \qquad g'(0) = \lambda^{2/3}$$

then

$$\frac{f^{(n)}(\infty)}{g^{(n)}(\infty)} = \frac{F^{(n)}(\infty)}{G^{(n)}(\infty)}$$

a result independent of λ.

Lastly, we determine a system of second order equations to which the method applies. Let us suppose the system has the form

$$\left.\begin{array}{l} \displaystyle\sum_{\substack{m,n,p,\\r,s,t,q}} A_{mnprstq} f''^{m} f'^{n} f^{p} g''^{r} g'^{s} g^{t} \eta^{q} = 0 \\[12pt] \displaystyle\sum_{\substack{\bar{m},\bar{n},\bar{p},\\\bar{r},\bar{s},\bar{t},\bar{q}}} A_{\bar{m}\bar{n}\bar{p}\bar{r}\bar{s}\bar{t}\bar{q}} f''^{\bar{m}} f'^{\bar{n}} f^{\bar{p}} g''^{\bar{r}} g'^{\bar{s}} g^{\bar{t}} \eta^{\bar{q}} = 0 \end{array}\right\} \tag{3.162}$$

having the boundary conditions

$$f(0) = 0, \quad f^{(d+1)}(\infty) = k, \quad g(0) = 0, \quad g^{(\bar{d}+1)}(\infty) = \bar{k}.$$

Here the indices, d and $\bar{d}$ and the A's are arbitrary constants, integers and coefficients, respectively. For independent μ and λ, let

$$f(\eta) = \lambda^{1-\alpha} F(\lambda^{\alpha}\eta), \qquad g(\eta) = \mu\lambda^{-\alpha} G(\lambda^{\alpha}\eta) \tag{3.163}$$

where $\lambda = f'(0)$, $\mu = g'(0)$ and $F(0) = 0$, $F'(0) = 1$, $G(0) = 0$, $G'(0) = 1$.

For *F* and *G* to satisfy Eq. (3.162) we find, upon substitution, that

$$\lambda^{(1+\alpha)m+n+(1-\alpha)p}\,\mu^{r+s+t}\,\lambda^{\alpha(r-t-q)} = c \quad \text{(const)}$$

$$\lambda^{(1+\alpha)\bar{m}+\bar{n}+(1-\alpha)\bar{p}}\,\mu^{\bar{r}+\bar{s}+\bar{t}}\,\lambda^{\alpha(\bar{r}-\bar{t}-\bar{q})} = \bar{c} \quad \text{(const)}.$$

If λ and μ are to be independent the four conditions

$$r + s + t = \text{const}, \qquad \bar{r} + \bar{s} + \bar{t} = \text{const}$$

$$(1+\alpha)m + n + (1-\alpha)p + \alpha(r - t - q) = \text{const}$$

$$(1+\alpha)\bar{m} + \bar{n} + (1-\alpha)\bar{p} + \alpha(\bar{r} - \bar{t} - \bar{q}) = \text{const}$$

must be satisfied. This restricts the form of Eq. (3.162).

The values of λ and μ are then given by

$$\begin{aligned}
\lambda &= \{k/F^{(d+1)}(\infty)\}^{1/(1+d\alpha)}, \\
\mu &= \bar{k}/\lambda^{d\alpha}G^{(d+1)}(\infty).
\end{aligned} \qquad (3.164)$$

Again we have reduced a boundary value problem to two initial value problems.

Extensions to similar systems of equations should follow this same type of analysis. Na [54] applied the theory of groups in transforming boundary value problems to initial value problems.

3.7 Nonlinear Equations from Diffusion

A variety of physical problems are modeled with the diffusion equation, where the diffusion coefficient depends upon the concentration,

$$\frac{\partial C}{\partial t} = \text{div}[D(C)\,\text{grad}\,C]. \qquad (3.165)$$

Among these we find, in addition to diffusion, heat conduction, fluid flow through porous media, boundary layer flow over a flat plate (von Mises form of the equations), solar prominences, and others. A discussion of the general uses of Eq. (3.165) is given in an article by Heaslet and Alksne [46]. Consideration of equations of this type with $D(C) = C^n$ have been carried out by Zel'dovic and Kompaneec [47] and Pattle [48] for source problems. Boyer [49] examines an extended problem in q dimension spherical similarity. His physical model, also of source type, is concerned with blasts. Crank [50] devotes an entire chapter to this subject aiming his remarks primarily at the ordinary differential equations obtained via the similarity transformation $\eta = x/t^{1/2}$ which he calls the Boltzmann transformation. That this may not be the natural similarity variable is pointed out by Ames [51] in his paper on the general similarity problem for the nonlinear diffusion equation.

The one dimensional form of Eq. (3.165) is

$$\frac{\partial C}{\partial t} = \frac{\partial}{\partial x}\left[D(C)\frac{\partial C}{\partial x}\right]. \tag{3.166}$$

In 1894 Boltzmann (see Crank [50]) discovered a similarity variable

$$\eta = x/t^{1/2}$$

for this equation. We find that

$$\frac{\partial C}{\partial x} = \frac{dC}{d\eta}\frac{\partial \eta}{\partial x} = \frac{1}{t^{1/2}}\frac{dC}{d\eta}$$

$$\frac{\partial C}{\partial t} = \frac{dC}{d\eta}\frac{\partial \eta}{\partial t} = -\frac{x}{2t^{3/2}}\frac{dC}{d\eta} = -\frac{\eta}{2t}\frac{dC}{d\eta}$$

and therefore

$$\frac{\partial}{\partial x}\left[D(C)\frac{\partial C}{\partial x}\right] = \frac{\partial}{\partial x}\left[\frac{D}{t^{1/2}}\frac{dC}{d\eta}\right] = \frac{1}{t}\frac{d}{d\eta}\left[D\frac{dC}{d\eta}\right].$$

Setting these into Eq. (3.166) we find the ordinary differential equation

$$\frac{d}{d\eta}\left[D(C)\frac{dC}{d\eta}\right] + \frac{1}{2}\eta\frac{dC}{d\eta} = 0. \tag{3.167}$$

As usual, for a similarity equation, the boundary conditions are $C = C_0$ at $\eta = 0$ and $C = C_1$ at $\eta = \infty$ or $C = C_1$ at $\eta = -\infty$, and $C = C_2$ at $\eta = \infty$.

We shall discuss some approximate methods for such equations as Eq. (3.167) in Chapter 4. Here we first discuss the work of Philip [52, 53] who determined a general method for establishing large classes of $D(C)$ functions which yield exact solutions of Eq. (3.167). Without loss of generality† we suppose the boundary conditions are

$$C = 1 \quad \text{when} \quad \eta = 0,$$
$$C = 0 \quad \text{when} \quad \eta = \infty. \tag{3.168}$$

Proceeding we rewrite Eq. (3.167) as

$$2d\left[D(C)\frac{dC}{d\eta}\right] = -\eta \, dC$$

and integrate from 0 to C—that is

$$\int_0^C \eta \, dC = -2D(C)\frac{dC}{d\eta}$$

or

$$D(C) = -\frac{1}{2}\frac{d\eta}{dC}\int_0^C \eta \, dC. \tag{3.169}$$

Therefore, the solution of Eq. (3.167) subject to Eq. (3.168) and $\eta(C)$ exists in exact form, so long as $D(C)$ is of the form

$$D = -\frac{1}{2}\frac{dF}{dC}\int_0^C F \, dC \tag{3.170}$$

where $\eta = F = F(C)$ is any single valued function on $0 \le C \le 1$. In addition F must satisfy the conditions to be imposed on η in Eq. (3.169). Thus

$$F(1) = 0. \tag{3.171}$$

For D to exist for all values of C in $0 \le C \le 1$, it is necessary that dF/dC and $\int_0^C F \, dC$ exist on this range.‡ Lastly D may be required to be positive. The flux Q of the diffusing substance in the positive x direction is

$$Q = -D\frac{\partial C}{\partial x} = -t^{-1/2}D\frac{dC}{d\eta}.$$

---

† Extensions to the slightly more general case with $C = C_0$ at $\eta = 0$ and $C = C_1$ at $\eta = \infty$ involves the transformation $\psi = (C - C_1)/(C_0 - C_1)$ to convert this problem to the one herein considered.

‡ If a finite number of discontinuities in D are allowable, or if D is permitted to be infinite at a finite number of points on $0 < C \le 1$, dF/dC may either not exist or be infinite at the appropriate number of points.

For $Q \geq 0$ we must have $dC/d\eta \leq 0$. Accordingly the further condition

$$\frac{dF}{dC} \leq 0 \quad \text{on} \quad 0 \leq C \leq 1 \tag{3.172}$$

must be imposed.

Exact solutions of Eq. (3.167) subject to Eq. (3.168) are assured by simply selecting F functions satisfying the above conditions. Thus if we select $F = 1 - C^n$, $n > 0$ we readily find via Eq. (3.170) that

$$D(C) = \tfrac{1}{2}nC^n[1 - C^n/(1 + n)].$$

Further examples and extensions of this idea are given by Philip [52].

Ames [51] considers the m-dimensional diffusion equation with spherical symmetry

$$\frac{1}{r^{m-1}} \frac{\partial}{\partial r}\left[r^{m-1} C^n \frac{\partial C}{\partial r} \right] = \frac{\partial C}{\partial t} \tag{3.173}$$

for source type boundary conditions

$$\begin{aligned}
C &= G(t) \quad \text{for} \quad r = 0, \quad t > 0 \\
C &= 0 \quad\quad \text{for} \quad t = 0, \quad r > 0 \\
C &= 0 \quad\quad \text{for} \quad r = \infty, \quad t > 0.
\end{aligned} \tag{3.174}$$

In this case the proper similarity variables are

$$\phi = r/t^A, \quad f(\phi) = \frac{C(r, t)}{t^\alpha}, \quad \alpha = \frac{2A - 1}{n} \tag{3.175}$$

where A is an arbitrary constant. The ordinary differential equation for f is

$$[\phi^{m-1} f^n f']' = \alpha \phi^{m-1} f - A\phi^m f' \tag{3.176}$$

where $'$ indicates differentiation with respect to ϕ.

Since A is a free parameter we choose it in such a way that a first integration of Eq. (3.176) is possible. For the right-hand side of Eq. (3.176) to be of the form $-(B\phi^m f)'$ requires that $B = A$ and $mA = (1 - 2A)/n$. This implies that if A is chosen as

$$A = (nm + 2)^{-1} \tag{3.177}$$

then Eq. (3.176) becomes

$$(\phi^{m-1} f^n f')' + (nm + 2)^{-1} (\phi^m f)' = 0,$$

which is integrable to

$$f^n f' + (nm + 2)^{-1} \phi f = \Gamma \phi^{1-m}. \tag{3.178}$$

Here Γ is an integration constant.

The boundary conditions derived from Eq. (3.174) are

$$f(\infty) = 0, \quad f(0) = t^{-\alpha} C(0, t) = t^{-\alpha} G(t) \tag{3.179}$$

and for these to be independent of t requires

$$g(t) = t^\alpha \text{ where } \alpha = -m/(nm + 2).$$

With this choice of $G(t)$ the boundary conditions on f become

$$f(0) = 1 \quad \text{and} \quad f(\infty) = 0. \tag{3.179a}$$

To these we adjoin the additional condition of vanishing flux —that is,

$$C^n \frac{\partial C}{\partial x} = 0 \quad \text{for} \quad t = 0, \quad \text{all} \quad r > 0. \tag{3.180}$$

Now we admit the possibility that concentration and flux vanish at a finite critical value (to be determined) of ϕ, say $\phi_0(n, m)$, and are both zero for $\phi_0 \leq \phi \leq \infty$. With this possibility Eq. (3.178) can be further transformed by means of

$$F(\phi) = f^{n+1}(\phi)$$

to

$$(n + 1)^{-1} F'(\phi) = \Gamma \phi^{1-m} - (nm + 2)^{-1} F^{1/(n+1)} \tag{3.181}$$

with boundary conditions

$$F(0) = 1, \quad F'(\phi_0) = F(\phi_0) = 0. \tag{3.181a}$$

The last two conditions imply $\Gamma = 0$. Lastly, an elementary integration gives

$$F(\phi) = \left\{ \frac{n}{2(nm + 2)} \left[\phi_0{}^2 - \phi^2 \right] \right\}^{(n+1)/n} \tag{3.182}$$

From $F(0) = 1$ we find the value of ϕ_0 to be

$$\phi_0 = \left[\frac{2(nm + 2)}{n} \right]^{1/2}. \tag{3.183}$$

As $n \to \infty$, $\phi_0 \to (2m)^{1/2}$, and as $n \to 0$, $\phi_0 \to \infty$. This exact dependence of ϕ_0 on n and m is qualitatively the same as the numerical result of Heaslet and Alksne [46] for a related problem.

The exact solution for $C(r, t)$ is obtained by composing Eqs. (3.175), (3.182), and (3.183) to

$$C(r, t) = t^{-m/(nm+2)} \left\{ \frac{n}{2(nm + 2)} \left[\phi_0{}^2 - \phi^2 \right] \right\}^{1/n}.$$

REFERENCES

1. Tong, K. N., "Theory of Mechanical Vibration," Wiley, New York, 1960.
2. Pipes, L. A., "Applied Mathematics for Engineers and Physicists," McGraw-Hill, New York, 1958.
3. Ames, W. F., *Ind. Eng. Chem. Fundamentals* **1,** 214 (1962).
4. Pipes, L. A., "Matrix Methods for Engineering," Prentice-Hall, Englewood Cliffs, New Jersey, 1963.
5. Albert, A. A., "Introduction to Algebraic Theories," Univ. of Chicago Press, Chicago, Illinois, 1940.
6. Hildebrand, F. B., "Methods of Applied Mathematics," Prentice-Hall, Englewood Cliffs, New Jersey, 1952.
7. Benson, S. W., "The Foundations of Chemical Kinetics," McGraw-Hill, New York, 1960.
8. Murdoch, P. G., *A.I.Ch.E. J.* **7,** 526 (1961).
9. Abraham, W. H., *Ind. Eng. Chem. Fundamentals* **2,** 221 (1963).
10. Jury, E. I., "Sampled Data Control Systems," Wiley, New York, 1958.
11. Ragazzini, J. R., and Franklin, G. F., "Sampled Data Control Systems," McGraw-Hill, New York, 1958.
12. Carrier, G. F., Krook, M., and Pearson, C. E., "Functions of a Complex Variable," McGraw-Hill, New York, 1966.
13. Churchill, R. V., "Complex Variables and Applications," McGraw-Hill, New York, 1960.
14. Bharucha-Reid, A. T., "Elements of the Theory of Markov Processes and Their Applications," McGraw-Hill, New York, 1960.
15. Kilkson, H., *Ind. Eng. Chem. Fundamentals* **3,** 281 (1964).
16. Flory, P. J., "Principles of Polymer Chemistry," Cornell Univ. Press, Ithaca, New York, 1953.
17. Moon, P., and Spencer, D. E., "Field Theory Handbook," Springer, Berlin, 1961.
18. Bateman, H., "Partial Differential Equations of Mathematical Physics," Cambridge Univ. Press, London and New York, 1959.

19. Richardson, O. W., "The Emission of Electricity from Hot Bodies," Longmans, Green, New York, 1921.
20. Walker, G. W., *Proc. Roy. Soc.* **A91**, 410 (1915).
21. Chambré, P. L., *J. Chem. Phys.* **20**, 1795 (1952).
22. Lempke, H., *J. Reine Angew. Math.* **142**, 118 (1913).
23. Emden, V. R., "Gaskugeln." Springer, Berlin, 1907.
24. Chandrasekhar, S., "Introduction to Study of Stellar Structure," Chicago Univ. Press, Chicago, Illinois, 1939.
25. Davis, H. T., "Introduction to Nonlinear Differential and Integral Equations," Dover, New York, 1962.
26. Miller, J. C. P., "Tables for Emden Functions." Brit. Assoc. Advance. Sci., London, 1932.
27. Hansen, A. G., "Similarity Analyses of Boundary Value Problems in Engineering," Prentice-Hall, Englewood Cliffs, New Jersey, 1964.
28. Ames, W. F., "Nonlinear Partial Differential Equations in Engineering," Academic Press, New York, 1965.
29. Kline, S. J., "Similitude and Approximation Theory," McGraw-Hill, New York, 1965.
30. Sedov, L. I., "Dimensional and Similarity Methods in Mechanics" (transl.). Academic Press, New York, 1960.
31. Bickley, W. G., *Phil. Mag.* **23**, 727 (1937).
32. Schlichting, H., "Boundary Layer Theory," 4th ed. (transl.), McGraw-Hill, New York, 1960.
33. Falkner, V. M., and Skan, S. W., *Phil. Mag.* [7] **12**, 865 (1931).
34. Pohlhausen, K., *Z. Angew. Math. Mech.* **1**, 252 (1921).
35. Schlichting, H., *Z. Angew. Math. Mech.* **13**, 260 (1933).
36. Ames, W. F., Similarity solutions for anisentropic gas flows, Unpublished lecture, Iowa State University, Ames, Iowa, 1966.
37. Kamke, E., "Differentialgleichungen, Lösungsmethoden und Lösungen," Vol. 1, Akad. Verlagsges., Leipzig, 1956.
38. Lock, G. S. H., *Bull. Mech. Eng. Educ.* **5**, 71 (1966).
39. Hildebrand, F. B., "Methods of Applied Mathematics," Prentice-Hall, Englewood Cliffs, New Jersey, 1952.
40. Weyl, H., *Ann. Math.* **43**, 381 (1942).
41. Homann, F., *Z. Angew. Math. Mech.* **16**, 153 (1936).
42. Siekmann, J., *Z. Angew. Math. Phys.* **13**, 183 (1962); see also Langer, R. E., "Nonlinear Problems," p. 298. Univ. of Wisconsin Press, Madison, Wisconsin, 1963.
43. Siekmann, J., *Z. Flugwiss.* **10**, 248 (1962).
44. Klamkin, M. S., *SIAM Rev.* **4**, 43 (1962).
45. Greenspan, H. P., and Carrier, G. F., *J. Fluid Mech.* **6**, 77 (1959).
46. Heaslet, M. A., and Alksne, A., *J. Soc. Ind. Appl. Math.* **9**, 584 (1961).
47. Zel'dovic, Ya. B., and Kompaneec, A. S., *Izv. Akad. Nauk* SSR 61, (1950).
48. Pattle, R. E., *Quart J. Mech. Appl. Math.* **12**, 407 (1959).

49. Boyer, R. H., *J. Math. and Phys.* **41,** 41 (1962).
50. Crank, J., "The Mathematics of Diffusion," Oxford Univ. Press, London and New York, 1956.
51. Ames W. F., *Ind. Eng. Chem. Fundamentals* **4,** 72 (1965).
52. Philip, J. R., *Australian J. Phys.* **13,** 1 (1960).
53. Philip, J. R., *Australian J. Phys.* **13,** 13 (1960).
54. Na, T. Y., *SIAM Rev.* **9,** 204 (1967).

4

APPROXIMATE METHODS

Introduction

By approximate methods we shall mean analytical procedures for developing solutions in the form of functions which are close, in some sense, to the exact, but usually unknown, solution of the nonlinear problem. Therefore numerical methods fall into a separate category (see Chapter 5) since they result in tables of values rather than functional forms.

Approximate methods may be divided into three broad interrelated categories; "iterative," "asymptotic," and "weighted residual." The iterative methods include the development of series, methods of successive approximation, rational approximations, and the like. Some form of repetitive calculation via some operation F whose character is $u_{n+1} = F[u_n, u_{n-1}, \ldots]$ successively improves the approximation. Transformation of the equation to an integral equation leads to a natural iterative method.

Asymptotic procedures have at their foundation a desire to develop solutions that are approximately valid when a physical parameter (or variable) of the problem is very small, very large or in close proximity to some characteristic value. Typical of these methods are the perturbation procedures both regular and singular.

The weighted residual methods, probably originating in the calculus of variations, require that the approximate solution be close to the exact solution in the sense that the difference between them (residual) is somehow minimized. Collocation insists that the residual vanish at a predetermined set of points while Galerkin's method is so formulated that weighted integrals of the residual vanish. These *error distribution* techniques are sometimes called *direct methods* of the calculus of variations although they need not be related to a variational problem.

Since these procedures are approximate an important question concerning the accuracy of approximation must be asked. In many cases convergence theorems exist to the effect that if an iteration is carried on indefinitely, or if an interval size is decreased without limit, then the process converges to the true solution. While these theorems are not without their stimulation value to the confidence of the analyst they are not as much practical value as a realistic error bound applicable at any stage of the computation.

Realistic error bounds or estimates exist for some approximate processes. Perhaps the best known of these is that for the Taylor series and the method of successive approximations for integral equations. Where these bounds are not available a common method of demonstrating the utility of an approximation process has been to apply it to a problem whose solution is known exactly. The error is then calculable. Then the presumption is that the method will produce errors of the same magnitude on similar problems. This is not an ideal procedure but one we must use on occasions.

4.1 Some Mathematical Properties

Equations of order n, $n > 1$ are reducible to n simultaneous first order equations. This reduction is *not unique* as will be demonstrated for the third order Blasius equation

$$w''' + ww'' = 0. \tag{4.1}$$

One obvious substitution is $u = w'$, $v = u'$ ($= w''$) so that Eq. (4.1) is equivalent to the three equations

$$\frac{dw}{d\eta} = u$$

$$\frac{du}{d\eta} = v \tag{4.2}$$

$$\frac{dv}{d\eta} = -wv.$$

An alternate to this procedure is to set (say) $u = w' + w$, $v = u'$ whereupon

Eq. (4.1) is equivalent to the three equations

$$\frac{dw}{d\eta} = u - w$$

$$\frac{du}{d\eta} = v \qquad (4.3)$$

$$\frac{dv}{d\eta} = v - u + w[1 - v + u - w].$$

Occasionally transformations other than the immediately obvious one that leads to Eqs. (4.2) may be more convenient. Such a situation occurred in the work of Liénard [1] in his investigation of limit cycles† for the general second order equation

$$x'' + f(x)x' + x = 0. \qquad (4.4)$$

Equation (4.4) may be written as

$$\frac{d}{dt}\left[\frac{dx}{dt} + F(x)\right] + x = 0$$

where $F'(x) = f(x)$ and instead of the previous substitutions Liénard writes $y = (dx/dt) + F(x)$ whereupon Eq. (4.4) is equivalent to the two first order equations

$$\frac{dx}{dt} = y - F(x), \qquad \frac{dy}{dt} = -x. \qquad (4.5)$$

The physical problems modeled by differential equations are of the two types, *propagation or initial value problems* and *equilibrium or boundary value problems*. In the initial value problem the analyst is faced with the prediction of the future behavior of the system given the initial data and the equation. In an equilibrium problem one seeks a configuration which satisfies the equation subject to side boundary conditions. Closely related to these questions are eigenvalue problems. Only the configuration is required in an equilibrium problem. But in an eigenvalue problem one needs to determine the value or values of a scalar parameter (eigenvalues) in addition to the configuration.

---

† For a discussion of limit cycles the reader is referred to Davies and James [2].

Some guiding principles in the form of existence and uniqueness theorems are available. These will be stated in the sequel without proof. The interested reader may find various proofs in the work of Davies and James [2], Goursat *et al.* [3], and Ince [4].

Our general *propagation problem* may be expressed as a set of governing differential equations

$$\frac{dy_j}{dt} = f_j(y_1, ..., y_n, t), \qquad j = 1, ..., n \qquad (4.6)$$

subject to the initial conditions $y_j(t_0) = y_{j0}$, for the variables $y_j(t)$. The course of a typical propagating variable in a "well behaved" system might be as shown in Fig. 4-1a. Since integration is a "smoothing" process we

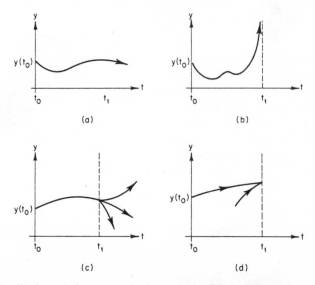

Fig. 4-1. Qualitative solution properties for ordinary differential equations. (a) Well-behaved solution. (b) Solution does not exist at t_1. (c) Solution not unique at t_1. (d) Nonuniqueness at t_1 and no solution for $t > t_1$.

can generally expect the $y_j(t)$ to be smoother functions than the f_j. Of course the solution may fail to exist or be nonunique. Some types of difficulties are displayed in Fig. 4-1b–d. The presence of such singularities is not always immediately obvious but their possible presence should always be suspected. Very often they can be detected by applying the tests given in the theorems below.

Cauchy's Existence—Uniqueness Theorem

If all f_j, $j = 1, 2, \ldots, n$ are single valued and continuous in a neighborhood D of the initial point $(y_{10}, y_{20}, \ldots, y_{n0}, t_0)$ and satisfy the n Lipschitz conditions

$$\left| f_j(\bar{y}_1, \ldots, \bar{y}_n, t) - f_j(y_1, \ldots, y_n, t) \right| < K \sum_{k=1}^{n} \left| \bar{y}_k - y_k \right| \qquad (4.7)$$

where K is some positive constant and $(\bar{y}_1, \ldots, \bar{y}_n, t)$ and $(y_1, \ldots, y_n, t)$ are distinct points of D, then there exists one and only one solution of Eqs. (4.6) reducing to y_{j0} at $t = t_0$ and continuous in D.

For many engineering problems this theorem is overly powerful. One may replace the Lipschitz condition, Eq. (4.7), by the requirement that all the partial derivatives $\partial f_j / \partial x_k$, $(j, k = 1, 2, \ldots, n)$ be finite continuous functions of all their arguments. This weaker theorem is not always applicable however. To illustrate such a case consider the problem of calculating the velocity u in a hydraulic surge tank (see McLachlan [5]). The appropriate differential equation is

$$\frac{d^2u}{dt^2} + 2k \left| u \right| \frac{du}{dt} + au = d \qquad (4.8)$$

where k, a, and d are physical parameters. Let us set $v = du/dt$ so that Eq. (4.8) becomes the two simultaneous equations

$$\frac{du}{dt} = v, \qquad \frac{dv}{dt} = d - 2k \left| u \right| v - au. \qquad (4.9)$$

During the surging the velocity u changes sign. But $\left| u \right|$ has no derivative at the origin. However, it does satisfy the Lipschitz condition.

A point $(y_1, y_2, \ldots, y_n, t)$ at which all the conditions of Cauchy's theorem are satisfied is called an *ordinary point* for the differential equation. Thus in the neighborhood of an ordinary point the theorem establishes the existence, uniqueness, and continuity of the solution of Eqs. (4.6).

A second theorem guarantees that the solution may be expanded in a Taylor series centered about an ordinary point.

Expansion Theorem

Let the f_j of Eqs. (4.6) have convergent $(n + 1)$ dimensional Taylor series in the domain D: $\left| y_j - \eta_j \right| < a, \left| t - t_0 \right| < b$ where $(\eta_1, ..., \eta_n, t_0)$ is an ordinary point. If in D, $\max_j \left| f_j \right| < M$, then the functions $y_j(t)$ have Taylor series in $t - t_0$ which converge for $\left| t - t_0 \right| < m = \min(b, M/a)$.

4.2 Series Expansions

In this section we develop procedures for the term by term construction of Taylor's series. Following this we describe how the series may be economized by using a class of orthogonal polynomials known as Chebyshev polynomials. The section is concluded by applying *analytic continuation* to extend the range of validity of the series expansion. In what follows we use the notation y_j' to indicate derivatives.

Taylor's Expansion

The Taylor's series expansion of a function $y_j(t)$ about the ordinary point $t = t_0$ is well known to be

$$y_j(t) = y_j(t_0) + \sum_{n=1}^{\infty} y_j^{(n)}(t_0) \frac{(t - t_0)^n}{n!}. \tag{4.10}$$

The development of this expansion requires only the value of the function and its derivatives at a single point. Practically, we only use a finite number of terms in Eq. (4.10)—let us say $N + 1$. Thus the series is *truncated* after the terms in $(t - t_0)^N$ introducing a *truncation error* whose value is found from the calculus (see Thomas [6] for example) to be

$$E_N(t, t_0) = y_j^{(N+1)}(\tau) \frac{(t - t_0)^{N+1}}{(N + 1)!} \tag{4.11}$$

where $t_0 < \tau < t$. We can in general only estimate the derivative in Eq. (4.11) since the function and the value of τ are not known exactly. But if we know that $m \le \left| y_j^{(N+1)}(t) \right| \le M$ for $t_0 \le t \le t_1$ then we know that

$$m \frac{(t_1 - t_0)^{N+1}}{(N+1)!} \leq |E_N(t_1, t_0)| \leq M \frac{(t_1 - t_0)^{N+1}}{(N+1)!}. \qquad (4.12)$$

The size of the error in Eq. (4.12) can be used to decide the number of terms required to yield a specified accuracy.

The development of the series for Eqs. (4.6), subject to the initial conditions $y_j(t_0)$, requires the calculation of the coefficients in Eq. (4.10). To obtain these we proceed as follows: The initial conditions provide the first terms, $y_j(t_0)$, directly. Upon substituting these values into the right hand sides of the governing Eqs. (4.6) with $t = t_0$ we obtain the values $y_j'(t_0)$. To obtain the second derivatives at $t = t_0$ we differentiate the Eqs. (4.6) and find

$$y_j'' = \frac{\partial f_j}{\partial t} + \sum_{k=1}^{n} \frac{\partial f_j}{\partial y_k} y_k', \qquad j = 1, 2, ..., n. \qquad (4.13)$$

At $t = t_0$ we already have $y_j(t_0), y_j'(t_0)$ so the right hand side of Eq. (4.13) is directly evaluable to give $y_j''(t_0)$. A continuation of this process allows us to develop as many terms of the series as desired.

As an illustration of the process consider the third order equation $w''' + ww''$, $w(0) = 0$, $w'(0) = 0$, $w''(0) = 1$ developed in Chapter 3, when boundary value problems were converted into initial value problems. With $t_0 = 0$ the first three terms are already available. Clearly $w'''(0) = 0$. To obtain the remainder we develop the next few derivatives:

$$w^{(4)} = -ww''' - w'w'', \qquad \text{hence } w^{(4)}(0) = 0;$$

$$w^{(5)} = -ww^{(4)} - 2w'w''' - (w'')^2, \qquad \text{hence } w^{(5)}(0) = -1;$$

$$w^{(6)} = -ww^{(5)} - 3w'w^{(4)} - 4w''w''', \qquad \text{hence } w^{(6)}(0) = 0;$$

$$w^{(7)} = -ww^{(6)} - 4w'w^{(5)} - 7w''w^{(4)} - 4(w''')^2, \qquad \text{hence } w^{(7)}(0) = 0;$$

$$w^{(8)} = -ww^{(7)} - 5w'w^{(6)} - 11w''w^{(5)} - 11w'''w^{(4)}, \text{hence } w^{(8)}(0) = +11.$$

The first few terms of the series expansion are

$$w(t) = \frac{t^2}{2!} - \frac{t^5}{5!} + \frac{11t^8}{8!} + E_8 \qquad (4.14)$$

where

$$|E_8(t_1, 0)| \leq \max_{0 \leq t \leq t_1} |w^{(9)}| \frac{t_1^{\,9}}{9!}. \qquad (4.15)$$

Evaluation of an upper bound for E_8 is difficult because it involves estimating the maximum value of a high order derivative of an unknown solution. One way of carrying out this estimation will now be illustrated. Let us estimate the truncation error of Eq. (4.14) on the interval $0 \leq t \leq 0.2$. According to Eq. (4.15) this requires the maximum value of the ninth derivative in this interval. Continuing our derivative computation we find $w^{(9)}(0) = 0$, $w^{(10)}(0) = 0$, and $w^{(11)}(0) = -375$. Thus the series reads

$$w(t) = \frac{t^2}{2!} - \frac{t^5}{5!} + \frac{11t^8}{8!} - \frac{375t^{11}}{11!} + \cdots. \tag{4.16}$$

From this we find

$$w^{(9)}(t) = -\frac{375t^2}{2!}, \qquad w^{(10)}(t) = -375t$$

and $\max_{0 \leq t \leq 0.2} \left| w^{(9)}(t) \right| = 375(0.04)/2 = 7.50$. We make the conservative estimate that the $\max_{0 \leq t \leq 0.2} \left| w^{(9)}(t) \right| < 20$. Accordingly we find the truncation error will be less than

$$20 \frac{(0.2)^9}{9!} = 1.99 \times 10^{-10}, \tag{4.17}$$

i.e., we are probably safe in expecting w to be correct to at least nine decimal places in $0 \leq t \leq 0.2$.

The accuracy of a numerical computation based upon a series depends in addition upon the adopted *round off* policy during the computation. We might for example evaluate each individual term to eleven decimal places, sum the terms and then round off to nine decimal places. Or we might round off each term separately to nine places before summing. As discussed by Hildebrand [7] these two results may differ by as much as P units in the ninth decimal place when the number of summed terms is $2P$.

Chebyshev Economization

A general discussion of orthogonal† polynomials is given by Hildebrand [7], Hamming [8], or Todd [9]. In the class of orthogonal functions, the

---

† A set of functions $\{f^n(x)\}$ are said to be orthogonal relative to the function $w(x)$ on $a \leq x \leq b$ if $\int_a^b w(x)f_i(x)f_j(x)\,dx = 0$ for $i \neq j$ and is not zero for $i = j$.

subclass of orthogonal polynomials has a number of special properties a few of which are:

(a) They satisfy a three term recurrence relation.
(b) They convert to power series from and are easily computed.
(c) Their zeros interlace each other.

Some of the classical orthogonal polynomials are listed below:

Legendre polynomials: $P_r(x) = \dfrac{1}{2^r r!} \dfrac{d^r}{dx^r}(x^2 - 1)^r, \qquad w(x) = 1,$

$-1 \le x \le 1, \qquad P_{r+1}(x) = \dfrac{2r+1}{r+1} x\, P_r(x) - \dfrac{r}{r+1} P_{r-1}(x).$

Laguerre polynomials: $L_r(x) = e^x \dfrac{d^r}{dx^r}(x^r e^{-x}), \qquad w(x) = e^{-x},$

$0 \le x < \infty, \qquad L_{r+1}(x) = (1 + 2r - x)L_r(x) - r^2 L_{r-1}(x).$

Hermite polynomials: $H_r(x) = (-1)^r e^{x^2} \dfrac{d^r}{dx^r}[e^{-x^2}], \qquad w(x) = e^{-x^2},$

$-\infty < x < \infty, \qquad H_{r+1}(x) = 2xH_r(x) - 2rH_{r-1}(x).$

Chebyshev polynomials†: $T_r(x) = \cos(r\,\cos^{-1} x), \qquad w(x) = (1-x^2)^{-1/2},$

$-1 \le x \le 1, \qquad T_{r+1}(x) = 2xT_r(x) - T_{r-1}(x). \hfill (4.18)$

While these and other orthogonal polynomials are useful in various fields of analysis and numerical analysis our interest herein lies in the Chebyshev polynomials. In addition to the previous properties we note that the Chebyshev polynomials also have the *equal-ripple* property of Fourier series. That is to say they have alternating maxima and minima of the same size. This follows from the fact that they are the Fourier functions cos $r\theta$ in the disguise of the transformation

$$\theta = \cos^{-1} x. \qquad (4.19)$$

In addition to the recurrence relation Eq. (4.18) we note that the identity

$$\cos(m + n)\theta + \cos(m - n)\theta = 2\cos m\theta \cos n\theta$$

---

† These prove useful in problems where errors near the ends of an interval $-1 \le x \le 1$ are significant.

gives us by substituting Eq. (4.19)

$$T_{m+n}(x) + T_{m-n}(x) = 2T_m(x)\, T_n(x). \tag{4.20}$$

Setting $m = n$, we have

$$T_{2n}(x) = 2T_n^2(x) - 1$$

which is useful for producing a single high order polynomial.

In Table 4.1 we list the first ten Chebyshev polynomials and x^n representations in terms of them.

TABLE 4.1

$T_0 = 1,$	$1 = T_0,$
$T_1 = x,$	$x = T_1$
$T_2 = 2x^2 - 1,$	$x^2 = \frac{1}{2}(T_0 + T_2)$
$T_3 = 4x^3 - 3x,$	$x^3 = \frac{1}{4}(3T_1 + T_3)$
$T_4 = 8x^4 - 8x^2 + 1,$	$x^4 = \frac{1}{8}(3T_0 + 4T_2 + T_4)$
$T_5 = 16x^5 - 20x^3 + 5x,$	$x^5 = \frac{1}{16}(10T_1 + 5T_3 + T_5)$
$T_6 = 32x^6 - 48x^4 + 18x^2 - 1,$	$x^6 = \frac{1}{32}(10T_0 + 15T_2 + 6T_4 + T_6)$
$T_7 = 64x^7 - 112x^5 + 56x^3 - 7x,$	$x^7 = \frac{1}{64}(35T_1 + 21T_3 + 7T_5 + T_7)$
$T_8 = 128x^8 - 256x^6 + 160x^4 - 32x^2 + 1,$	$x^8 = \frac{1}{128}(35T_0 + 56T_2 + 28T_4 + 8T_6 + T_8)$
$T_9 = 256x^9 - 576x^7 + 432x^5 - 120x^3 + 9x,$	$x^9 = \frac{1}{256}(126T_1 + 84T_3 + 36T_5 + 9T_7 + T_9)$

Additional polynomials and tables of them are available in the National Bureau of Standards tables [10].

Chebyshev showed that of all polynomials $p_n(x)$ of degree n, with leading coefficient 1, the polynomial $T_n(x)/2^{n-1}$ has the smallest least upper bound for its absolute value in $-1 \leq x \leq 1$. Since the upper bound of $|T_n(x)|$ is 1,† then this upper bound is $1/2^{n-1}$. This property is of considerable interest in numerical computation. The expression "Chebyshev approximation" is interpreted to mean those approximations which attempt to keep the maximum error to a minimum.‡ This is in sharp contrast to least-squares approximation which keeps the average square error down but isolated extreme errors are permitted. The Chebyshev method allows a larger average square error but keeps the extreme errors down.

---

† This follows from the definition $T_n(x) = \cos(n \cos^{-1} x)$.

‡ This is sometimes called the *minimax principle*.

A related application of the Chebyshev approach is that of *economization of series* due primarily to Lanczos [10, 11]. Suppose that we have the truncated power series, with error,

$$f(x) = \sum_{k=0}^{n} A_k x^k + E_n(x) \qquad (4.21)$$

where it is known that $|E_n(x)| < \varepsilon_1 \, (-1 \leq x \leq 1)$ and that $\varepsilon_1 < \varepsilon$, where ε is the prescribed allowable error. In addition suppose that $|A_n| + \varepsilon_1$ is not a tolerable error so that we cannot safely neglect the last term in the approximation $\sum_{k=0}^{n} A_k x^k$. Of course, for a power series the error tends to be large at the ends of the interval and small in the middle going to zero at the expansion center. Since the Chebyshev polynomials are developed with large errors at the interval ends in mind it seems appropriate to expand (really a replacement from Table 4.1) the approximation in a series of Chebyshev polynomials. Thus

$$\sum_{k=0}^{n} A_k x^k = \sum_{k=0}^{n} a_k T_k(x). \qquad (4.22)$$

From Table 4.1 and Eq. (4.18) we note that the terms of highest degree in $T_k(x)$ are given by

$$T_k(x) = 2^{k-1} \left[x^k - \frac{k}{4} x^{k-2} + \cdots \right]. \qquad (4.23)$$

In view of Eq. (4.23) let us now express both sides of Eq. (4.22) in descending powers of x

$$A_n x^n + A_{n-1} x^{n-1} + \cdots = 2^{n-1} a_n \left(x^n - \frac{n}{4} x^{n-2} + \cdots \right)$$

$$+ 2^{n-2} a_{n-1} \left(x^{n-1} - \frac{n-1}{4} x^{n-3} + \cdots \right) + \cdots . \qquad (4.24)$$

For this to be an identity there follows

$$a_n = 2^{-(n-1)} A_n,$$
$$a_{n-1} = 2^{-(n-2)} A_{n-1}, \qquad (4.25)$$
$$a_{n-2} = 2^{-(n-3)} \left(A_{n-2} + \frac{n}{4} A_n \right), \cdots .$$

Consequently, for sufficiently large n, the coefficients of $T_n, T_{n-1}, \ldots,$ T_{n-m+1} in Eq. (4.22) will be small relative to the respective coefficients of $x^n, x^{n-1}, \ldots, x^{n-m+1}$ for some $m > 0$. Since $\left| T_k(x) \right| \leq 1$ for $-1 \leq x \leq 1$, it may happen that

$$(\left| a_{n-m+1} \right| + \cdots + \left| a_n \right|) + \varepsilon_1 < \varepsilon, \tag{4.26}$$

that is, the last m terms in the right hand member of Eq. (4.22) are negligible. The approximation, to the required tolerance, can then be written

$$f(x) \approx \sum_{k=0}^{n-m} a_k \, T_k(x) \tag{4.27}$$

after which Eq. (4.27) can be transformed to power series form

$$f(x) \approx \sum_{k=0}^{n-m} \bar{a}_k \, x^k. \tag{4.28}$$

If we have a truncated power series and convert to a Chebyshev expansion generally we can obtain a much lower order polynomial approximation by discarding the later Chebyshev terms. The error is not greatly increased by this process. Further, the process can be applied to *any* polynomial once the interval of interest $a \leq x \leq b$ is transformed by a linear change of variable to $-1 \leq x \leq 1$.

As an example of the process consider

$$y = 1 - \frac{t^2}{2} + \frac{t^3}{3} - \frac{t^4}{8} - \frac{t^5}{60} + \frac{47t^6}{720} + \cdots \tag{4.29}$$

which constitutes the series expansion for y in the unidirectional surging of the surge tank problem $w' = y$, $y' = -w - y^2$, $w(0) = -1$, $y(0) = +1$. Let us suppose that we wish our truncation error to be less than 10^{-4} on $0 \leq t \leq 0.2$. If we truncate the series after the term in t^3 the truncation error is smaller than 3×10^{-4}. If we truncate after the term in t^4 the truncation error is less than 4×10^{-5}. To apply economization we must transform the interval $0 \leq t \leq 0.2$ to $0 \leq x \leq 1$ (really $-0.2 \leq t \leq 0.2$ to $-1 \leq x \leq 1$ but only the positive range has physical meaning) by means of the linear transformation $x = 5t$. Thus Eq. (4.29) becomes

$$y \approx 1 - \frac{x^2}{2 \cdot 5^2} + \frac{x^3}{3 \cdot 5^3} - \frac{x^4}{8 \cdot 5^4} \tag{4.30}$$

where we must truncate *after* the x^4 term to achieve the desired accuracy. Using Table 4.1 we find

$$y \approx T_0 - \frac{1}{100}(T_0 + T_2) + \frac{1}{1500}(3T_1 + T_3)$$

$$- \frac{1}{40,000}(3T_0 + 4T_2 + T_4)$$

$$\approx \frac{39,597}{40,000}T_0 + \frac{1}{500}T_1 - \frac{101}{10,000}T_2 + \frac{1}{1500}T_3 - \frac{1}{40,000}T_4 . \quad (4.31)$$

In the interval $0 \leq x \leq 1$, discarding the last term would produce a change of less than 2.5×10^{-5}. The total error on $0 \leq x \leq 1$ without the term in T_4 would then be less than 6.5×10^{-5}, well within tolerance.

Converting Eq. (4.31) back to power series form we find via Table 4.1 that

$$y \approx \frac{40,001}{40,000} - \frac{101}{200}t^2 + \frac{t^3}{3} \quad (4.32)$$

satisfies our requirements. This polynomial of degree three closely resembles Eq. (4.29) truncated after the term in t^3. It is not as dramatic a change as occurs in some examples (see e.g., Hamming [8].) This is probably due to the conservatism of our truncation error estimates.

Analytic Continuation

From the theory of Taylor's series we know that a few terms provide high accuracy near the expansion center but as the distance from the center increases the accuracy rapidly decreases. To maintain a specified accuracy we can use more and more terms of the series or we can use the process of *analytic continuation*. In brief this process involves the use of several expansion centers dispersed through an interval as shown in Fig. 4-2.

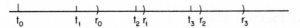

Fig. 4-2. Expansion centers and radii of convergence for analytic continuation.

The initial center at t_0 is used to develop a truncated series which has the desired accuracy on the interval $t_0 - r_0 \leq t \leq t_0 + r_0$. Within this interval, say at t_1, a second expansion is constructed using the values of the previous series at t_1 as the appropriate initial conditions. In this way the solution will consist of a sequence of series expansions each of which develops high accuracy near its expansion center. These series with different expansion centers represent the same function in overlapping intervals and provide what are called *analytic continuations* of the function.

There is a successive loss of accuracy as one continues this process. A rule of thumb is a loss of one significant figure for each separate expansion but obviously this depends upon the intervals between centers and the conservatism of the truncation error estimates.

4.3 Methods of Iteration

Approximate methods which generate a new estimate at the nth step in terms of one or more of the approximations in the preceding steps are called *methods of iteration*. Some of these methods have been extensively used by mathematicians in the development of existence theorems. Perhaps the best known of these iterative methods is the *Cauchy–Picard process*. Of less notoriety are some modifications of this process and those operational methods of Pipes [12, 13]. Strictly speaking perturbation methods are also iterative in nature but we shall discuss them under a separate heading.

Let L be a differential operator and

$$Lu = f. \tag{4.33}$$

If we wish to develop an iterative solution for Eq. (4.33) what is required is a *sequence* $u_0, u_1, ..., u_n, ...$ which *converges* to the exact solution as n increases indefinitely. Each iteration is based upon some relation of the form

$$u_{n+1} = F(u_n, u_{n-1}, ...) \tag{4.34}$$

where F is an operator intimately related to the equation to be solved.

CAUCHY–PICARD ITERATION

Consider the set of n simultaneous differential equations

$$\frac{dy_j}{dt} = f_j(y_1, ..., y_n, t), \qquad j = 1, 2, ..., n \tag{4.35}$$

having the prescribed initial values $y_j(t_0)$. Upon integrating from t_0 to t and evaluating the integration constants we have the relation

$$y_j(t) = y_j(t_0) + \int_{t_0}^{t} f_j(y_1, \ldots, y_n, t) \, dt. \tag{4.36}$$

The iteration process begins with a set of initial trial functions $y_j^{(0)}$ which may be arbitrarily chosen but usually are selected as the initial condition constants. A sequence of iterative approximations are then constructed according to the scheme† ‡

$$y_j^{(k+1)} = y_j(t_0) + \int_{t_0}^{t} f_j(y_1^{(k)}, y_2^{(k)}, \ldots, y_n^{(k)}, t) \, dt \tag{4.37}$$

$k = 0, 1, 2, \ldots; j = 1, 2, \ldots, n$. The following observations concerning Cauchy–Picard iteration are worthy of note:

(a) If the system (4.35) satisfies the assumptions of the existence-uniqueness theorem of Section 4.1 in some neighborhood of the initial data—then the Cauchy–Picard iteration converges to the true solution. This result was established by Picard [13a].

(b) This iterative process constructs the power series expansion (if it exists) which agrees with the Taylor series. If the Taylor series does not exist the iterative method can still be applied. Nonexistence of the Taylor series occurs if the expansion center is a singular point.

(c) Integrations are required. If the f_j are sufficiently complicated, analytic integration may be difficult. In such cases numerical or approximate integration may be necessary.

(d) This method is equivalent to replacing Eq. (4.35) by

$$\frac{dy_j^{(k+1)}}{dt} = f_j(y_1^{(k)}, \ldots, y_n^{(k)}, t),$$

$$y_j^{(k+1)}(t_0) = y_j(t_0). \tag{4.38}$$

The Cauchy-Picard iteration develops a sequence $\{y_j^{(k)}\}$ which converges to the true solution under the proper conditions. Often the rate of con-

---

† Note that the operator F of Eq. (4.34) is the integral of Eq. (4.37).

‡ The superscript index (k) in the notation $y_j^{(k)}$ indicates the step in the iteration and *not* the kth derivative.

vergence of this sequence can be *accelerated* by a slight modification of the method. We first illustrate with a simple example and then state and partially prove a general theorem of some utility in problems of fluid mechanics.

Consider the equation

$$\frac{dy}{dt} = y^2 + y, \qquad y(0) = 1/4 \tag{4.39}$$

whose exact solution is $y = e^t/(5 - e^t)$ and as $t \to 1$ $y \to e/(5 - e) \approx +1.191$. Using the Cauchy–Picard method

$$y^{(k)} = \tfrac{1}{4} + \int_0^t \{[y^{(k-1)}]^2 + y^{(k-1)}\} \, dt$$

we have the following early terms

$$y^{(0)} = \frac{1}{4},$$

$$y^{(1)} = \frac{1}{4} + \frac{5t}{16},$$

$$y^{(2)} = \frac{1}{4} + \frac{5t}{16} + \frac{15}{64} t^2 + \frac{25t^3}{768}.$$

Using the second term of the sequence we find that $y^{(0)}(1) = 0.250$, $y^{(1)}(1) = 0.562$ and $y^{(2)}(1) = 0.829$.

The preceding function sequence $\{y^{(k)}\}$ is obtained by replacing Eq. (4.39) by

$$\frac{dy^{(k)}}{dt} = y^{(k-1)} [y^{(k-1)} + 1], \qquad y^{(k)}(0) = \frac{1}{4}.$$

As an alternate suppose we seek a new sequence $y^{(k)}$ which satisfies

$$\frac{dy^{(k)}}{dt} = y^{(k)} [y^{(k-1)} + 1], \qquad y^{(k)}(0) = \frac{1}{4} \tag{4.40}$$

or its equivalent integral form

$$y^{(k)}(t) = y^{(k)}(0) \exp\left[\int_0^t (y^{(k-1)} + 1) \, dt\right]. \tag{4.41}$$

The early terms in the function sequence are therefore

$$y^{(0)} = \tfrac{1}{4},$$

$$y^{(1)} = \tfrac{1}{4}\exp[\tfrac{5}{4}\,t],$$

$$y^{(2)}(t) = \tfrac{1}{4}\exp\{\tfrac{5}{16}[\exp(\tfrac{5}{4}\,t) - 1] + t\}.$$

Thus $y^{(0)}(1) \approx 0.250$, $y^{(1)}(1) \approx 0.873$, and $y^{(2)}(1) \approx 1.483$.

The true value of $y(1) = 1.191$. At $t = 1$ the Cauchy–Picard method yields $y^{(2)}(1) = 0.829$ while for the modified method both $y^{(1)}(1)$ and $y^{(2)}(1)$ are more accurate. Note that $y^{(1)}(1)$ is smaller and $y^{(2)}(1)$ is larger than the true value. The process appears to be convergent since $|y^{(1)} - 1.191| = 0.318$ and $|y^{(2)}(1) - 1.191| = 0.292$. *Oscillating convergence is one of the fine features of the modification.*

We now consider the general problem from which the preceding example comes. For simplicity this problem is analyzed in terms of the single equation

$$\frac{dy}{dt} = G(t, y), \qquad y(t_0) = \alpha > 0, \qquad (4.42)$$

but the results are easily generalized to systems such as Eq. (4.35). The convergence of a sequence $\{y^{(k)}\}$ which satisfies the recursive equation

$$\frac{dy^{(k)}}{dt} = -y^{(k)} f(t, y^{(k-1)})$$

$$y^{(k)}(t_0) = \alpha > 0, \qquad k \geq 2 \qquad (4.43)$$

is to be examined. If a limit function $y^* = \lim_{k \to \infty} y^{(k)}$ exists then it is a solution of the equation

$$\frac{dy}{dt} = -yf(t, y) = G(t, y). \qquad (4.44)$$

The answer is given by the following theorem:

If in the interval I: $t_0 < t < t_0 + h(h > 0)$, the function $y(t)$ is positive and $f(t, y)$ has the properties:
  (a) $0 < f(t, y) < m$ in I,
  (b) $f(t, y)$ *is a monotone increasing function of y*,
  (c) $f(t, Y) - f(t, y) < K(Y - y)$,† $K > 0$ *then the sequence of positive*

---

† This is similar to but not identical to a Lipschitz condition.

functions $\{y^{(k)}\}$ *satisfying Eq.* (4.43) *converges to a limit function* y^* *satisfying Eq.* (4.44).

A complete proof of this theorem is given by Davies and James [2], for example. We note here that the convergence of the sequence $\{y^{(k)}\}$ is oscillatory in the sense established below.

From Eq. (4.43) we see that the sequence $\{y^{(k)}\}$ will satisfy the iterative equation

$$y^{(k)} = \alpha \exp\left[-\int_{t_0}^{t} f\{v, y^{(k-1)}(v)\} \, dv \right] = \alpha F(y^{(k-1)}). \qquad (4.45)$$

Now $f(t, y)$ is a positive monotone increasing function in I. Consequently, if $g(t) > g^*(t)$ at all points of I it follows that $F(g) < F(g^*)$ everywhere in I. Thus if $y^{(k-1)} < y^{(k)}$ then $F(y^{(k-1)}) > F(y^{(k)})$ whereupon $y^{(k)} > y^{(k+1)}$ since

$$y^{(k+1)} = \alpha F(y^{(k)}) < \alpha F(y^{(k-1)}) = y^{(k)}. \qquad (4.46)$$

Similarily if $y^{(k-1)} > y^{(k)}$ then $y^{(k)} < y^{(k+1)}$.

Now let $y^{(1)} = 0$. From Eq. (4.45) it is clear that $y^{(2)}$ is positive so that $y^{(1)} < y^{(2)}$. By the above argument it is then clear that

$$y^{(1)} < y^{(2)}, \qquad y^{(2)} > y^{(3)}, \qquad y^{(3)} < y^{(4)}, \qquad \dots . \qquad (4.47)$$

Since $y^{(3)}$ is positive $y^{(1)} < y^{(3)}$ and we deduce that

$$y^{(1)} < y^{(3)}, \quad y^{(2)} > y^{(4)}, \quad y^{(3)} < y^{(5)}, \quad y^{(4)} > y^{(6)}, \quad \dots . \qquad (4.48)$$

Thus the even functions form a decreasing sequence $y^{(2)} > y^{(4)} > y^{(6)} > \dots$ and the odd functions form an increasing sequence $y^{(1)} < y^{(3)} < y^{(5)} < \dots.$†
The former sequence will be bounded below and therefore has a limit $y_1^*(t)$. The latter is bounded above and therefore has a limit $y_2^*(t)$. The remainder of the proof establishes that in fact

$$y_1^* = y_2^*. \qquad (4.49)$$

4.4 Operational Iterative Methods

Nonlinear equations are often subdivided into two major types *autonomous* and *nonautonomous*. Autonomous equations are characterized by

---

† In these statements the assertions for even and odd sequences must be reversed if we begin counting from zero.

the fact that the independent variable does not appear explicitly in the equation. In many technical problems concerning nonlinear systems a "forcing function" appears. Thus the mathematical model is a differential equation of nonautonomous type. The use of operational methods, based upon the Laplace transform, for solving both classes of problems has been extensively studied by Pipes [12, 13]. The introduction of operational processes brings to bear extensive transform tables thereby reducing the algebraic labor involved in obtaining practical solutions.

To introduce one of several operational adaptations to iterative processes we consider the equation

$$x'' + \omega^2 x + \lambda x^2 = A(t) \tag{4.50}$$

used in the theory of seismic waves by Nagaoka [14]. The effect of the joint action of two simple harmonic forces will be investigated by setting

$$A(t) = a_1 \cos \omega_1 t + a_2 \cos \omega_2 t \tag{4.51}$$

and by assuming the system has the initial conditions

$$x(0) = a, \qquad x'(0) = 0. \tag{4.52}$$

In the sequel the p − multiplied Laplace transform

$$y(p) = p \int_0^\infty e^{-pt} x(t)\, dt = Lx \tag{4.53}$$

will be used.

To solve Eq. (4.50) we let x take the representation

$$x(t) = \sum_{i=0}^{\infty} x_i(t) \tag{4.54}$$

where the x_i, whose transforms we denote by $y_i(p)$, are to be determined. If Eq. (4.54) is squared we have

$$x^2 = x_0{}^2 + 2x_0 x_1 + x_1{}^2 + 2(x_0 + x_1)x_2 + x_2{}^2 + \cdots. \tag{4.55}$$

The Laplace transform of Eq. (4.50) is

$$y = \frac{p^2 a}{(p^2 + \omega^2)} + \frac{p^2 a_1}{(p^2 + \omega^2)(p^2 + \omega_1{}^2)}$$

$$+ \frac{p^2 a_2}{(p^2 + \omega^2)(p^2 + \omega_2{}^2)} - \frac{\lambda}{(p^2 + \omega^2)} Lx^2. \tag{4.56}$$

The transform of Eq. (4.54) is

$$y = \sum_{i=0}^{\infty} y_i(p). \qquad (4.57)$$

Upon substituting Eq. (4.55) and Eq. (4.57) into Eq. (4.56) we find

$$\sum_{i=1}^{\infty} y_i(p) = \frac{p^2 a}{(p^2 + \omega^2)} + \frac{p^2 a_1}{(p^2 + \omega^2)(p^2 + \omega_1^2)}$$

$$+ \frac{p^2 a_2}{(p^2 + \omega^2)(p^2 + \omega_2^2)} - \frac{\lambda}{(p^2 + \omega^2)}$$

$$\times L[x_0^2 + 2x_0 x_1 + x_1^2 + 2(x_0 + x_1)x_2 + \cdots].$$

Now let

$$y_0 = \frac{p^2 a}{(p^2 + \omega^2)} + \frac{p^2 a_1}{(p^2 + \omega^2)(p^2 + \omega_1^2)} + \frac{p^2 a_2}{(p^2 + \omega^2)(p^2 + \omega_2^2)}$$

$$= Lx_0 \qquad (4.58)$$

$$y_1 = -\frac{\lambda}{(p^2 + \omega^2)} Lx_0^2 = Lx_1 \qquad (4.59)$$

$$y_2 = -\frac{\lambda}{p^2 + \omega^2} L(2x_0 x_1 + x_1^2) = Lx_2 \qquad (4.60)$$

$$y_3 = -\frac{\lambda}{p^2 + \omega^2} L[2(x_0 + x_1)x_2 + x_2^2] = Lx_3, \qquad (4.61)$$

and so forth. With the y_i's thus identified the system has the solution

$$x = \sum_{i=0}^{\infty} x_i(t) = L^{-1}\left[\sum_{i=0}^{\infty} y_i(p)\right]. \qquad (4.62)$$

If λ is less than ω^2 it has been shown by Poincaré [15] and Shaefer [16] under otherwise general conditions that this series converges. The solution Eq. (4.62) is the sum of several functions—the first approximation $x_0(t)$ is the solution of the linear system with $\lambda = 0$. Inverting Eq. (4.58) we have

$$x_0(t) = A \cos \omega t + A_1 \cos \omega_1 t + A_2 \cos \omega_2 t \qquad (4.63)$$

TABLE 4.2[a]

1. $L \sin \omega t = \omega p T(\omega)$

2. $L \cos \omega t = p^2 T(\omega)$

3. $L \sin^2 \omega t = \frac{1}{2}[1 - p^2 T(2\omega)]$

4. $L \sin^3 \omega t = \frac{1}{4}[3p\omega T(\omega) - 3\omega p T(3\omega)]$.

5. $L \cos^2 \omega t = \frac{1}{2}[1 + p^2 T(2\omega)]$

6. $L \cos^3 \omega t = \frac{1}{4}[3p^2 T(\omega) + p^2 T(3\omega)]$

7. $L \sin At \cos Bt = \frac{1}{2}[(A + B)pT(A + B) + (A - B)pT(A - B)]$

8. $L \cos At \cos Bt = \dfrac{p^2}{2}[T(A + B) + T(A - B)]$

9. $L \sin At \sin Bt = \dfrac{p^2}{2}[T(A - B) - T(A + B)]$

10. $L \sin At \sin Bt \sin Ct$

$$= \frac{p}{4}[(A + B - C)T(A + B - C) + (B + C - A)T(B + C - A)$$
$$+ (C + A - B)T(C + A - B) - (A + B + C)T(A + B + C)]$$

11. $L \sin At \cos Bt \cos Ct$

$$= \frac{p}{4}[(A + B - C)T(A + B - C) - (B + C - A)T(B + C - A)$$
$$+ (C + A - B)T(C + A - B) + (A + B + C)T(A + B + C)]$$

12. $L \sin At \sin Bt \cos Ct$

$$= \frac{p^2}{4}[T(B + C - A) - T(A + B - C)$$
$$+ T(C + A - B) - T(A + B + C)]$$

13. $L \cos At \cos Bt \cos Ct$

$$= \frac{p^2}{4}[T(A + B - C) + T(B + C - A)$$
$$+ T(C + A - B) + T(A + B + C)]$$

14. $L^{-1}p^2 T(a)T(b) = (\cos at - \cos bt)/(b^2 - a^2), \quad a \neq b$

15. $L^{-1}p^2 T^2(a) = t \sin at/2a$

16. $L^{-1}p T^2(a) = \dfrac{\sin at}{2a^3} - \dfrac{t \cos at}{2a^2}$

17. $L^{-1}T(a) = (1 - \cos at)/a^2$

18. $T(a)T(b) = [T(a) - T(b)]/(b^2 - a^2), \quad a \neq b$

---

[a] Adapted by permission from Pipes [12].

where

$$A = a + \frac{a_1}{\omega_1{}^2 - \omega^2} + \frac{a_2}{\omega_2{}^2 - \omega^2}.$$

$$A_1 = \frac{a_1}{\omega^2 - \omega_1{}^2}, \qquad A_2 = \frac{a_2}{\omega^2 - \omega_1{}^2} \qquad \omega_1 \neq \omega, \quad \omega_2 \neq \omega.$$

With the goal in mind of some simplifying mechanism let us now set

$$T(\omega) = \frac{1}{p^2 + \omega^2}. \tag{4.64}$$

A brief collection of these transforms is given in Table 4.2.

If $x_0{}^2$ is substituted into Eq. (4.59) and use is made of the above T transform we find that y_1 takes the form

$$y_1 = -\lambda T(\omega) \left\{ \frac{A^2}{2} [p^2 T(2\omega) + 1] + \frac{A_1{}^2}{2} [p^2 T(2\omega_1) + 1] \right.$$

$$+ \frac{A_2{}^2}{2} [p^2 T(2\omega_2) + 1] + p^2 A A_1 [T(\omega + \omega_1) + T(\omega - \omega_1)]$$

$$+ p^2 A A_2 [T(\omega + \omega_2) + T(\omega - \omega_2)]$$

$$\left. + p^2 A_1 A_2 [T(\omega_1 + \omega_2) + T(\omega_1 - \omega_2)] \right\} \tag{4.65}$$

The inverse transform of Eq. (4.65) is computed by means of Table 4.2. Note that considerable simplification results if entry 18 of Table 4.2 is used early in the inversion. Thus we find x_1 to be

$$x_1 = -\frac{\lambda}{2\omega^2} (A^2 + A_1{}^2 + A_2{}^2)$$

$$+ \cos \omega t \left\{ \frac{A^2}{3\omega^2} + \frac{A_1{}^2 + A_2{}^2}{2\omega^2} + \frac{A_1{}^2}{(2\omega^2 - 8\omega_1{}^2)} + \frac{A_2{}^2}{(2\omega^2 - 8\omega_2{}^2)} \right.$$

$$- \frac{A A_1}{(\omega_1{}^2 + 2\omega\omega_1)} - \frac{A A_1}{(\omega_1{}^2 - 2\omega\omega_1)} - \frac{A A_2}{(\omega_2{}^2 + 2\omega\omega_2)}$$

$$- \frac{A A_2}{(\omega_2{}^2 - 2\omega\omega_2)} + \frac{A_1 A_2}{[\omega^2 - (\omega_1 + \omega_2)^2]}$$

$$\left. + \frac{A_1 A_2}{[\omega^2 - (\omega_1 - \omega_2)^2]} \right\}$$

$$+ \frac{A^2}{6\omega^2} \cos 2\omega t + \frac{A_1{}^2}{(8\omega_1{}^2 - 2\omega^2)} \cos 2\omega_1 t$$

$$+ \frac{A_2{}^2}{(8\omega_2{}^2 - 2\omega^2)} \cos 2\omega_2 t + \frac{AA_1}{\omega_1{}^2 + 2\omega\omega_1} \cos(\omega + \omega_1)t$$

$$+ \frac{AA_1}{(\omega_1{}^2 - 2\omega\omega_1)} \cos(\omega - \omega_1)t + \frac{AA_2}{\omega_2{}^2 + 2\omega\omega_2} \cos(\omega + \omega_2)t$$

$$+ \frac{AA_2}{\omega_2{}^2 - 2\omega\omega_2} \cos(\omega - \omega_2)t + \frac{A_1 A_2}{[(\omega_1 + \omega_2)^2 - \omega^2]}$$

$$\times \cos(\omega_1 + \omega_2)t + \frac{A_1 A_2}{[(\omega_1 - \omega_2)^2 - \omega^2]} \cos(\omega_1 - \omega_2)t.$$

A second correction x_2 is obtained from Eq. (4.60) in a like manner. It will contain λ^2 as a coefficient and may be neglected for $\lambda \ll \omega^2$.

The response of linear oscillators differs considerably from the form $x = x_0 + x_1$. The differences are that x undergoes a constant displacement $(-\lambda/2\omega^2)(A^2 + A_1{}^2 + A_2{}^2)$ from the static equilibrium $x = 0$; overtones of frequency 2ω, $2\omega_1$, $2\omega_2$ and combination tones of frequencies $\omega + \omega_1$, $\omega + \omega_2$, $\omega_1 + \omega_2$, $\omega - \omega_1$, $\omega - \omega_2$, and $\omega_1 - \omega_2$ are present.

This method of analysis may also be applied to obtain solutions to equations of the type

$$x'' + kx' + hx + ax^2 + bxx' + c(x')^2 = A(t).$$

Yet another iterative operational method, devised by Pipes, is intimately related to Lalesco's nonlinear integral equation (see Volterra [17]). We illustrate the procedure by using the second order nonlinear equation in operator form†

$$Z(D)x + f(x) = e(t) \tag{4.66}$$

where $D = d/dt$. The term $Z(D)$ is a linear operator. Let us suppose that $x(0) = x'(0) = 0$ although the problem is analyzable in the case of nonzero initial conditions. Again we use the p − multiplied Laplace transform Eq. (4.53) and set

$$Lx(t) = y(p), \qquad Le(t) = E(p), \qquad Lf[x(t)] = G(p). \tag{4.67}$$

---

† By $f(x)$ we mean an operator on x. Thus $f(x) = x^2 + (d/dt)(x^3)$ is allowable.

With this notation and initial conditions Eq. (4.66) is transformed into

$$Z(p)\,y(p) = E(p) - G(p) \tag{4.68}$$

or

$$y(p) = \frac{E(p)}{Z(p)} - \frac{G(p)}{Z(p)} \tag{4.69}$$

Let

$$H(p) = p/Z(p) \tag{4.70}$$

so that Eq. (4.69) may be rewritten as

$$y(p) = \frac{E(p)\,H(p)}{p} - \frac{G(p)\,H(p)}{p}. \tag{4.71}$$

If we now set $L^{-1}H = L^{-1}p/Z(p) = h(t)$ then $x(t)$ is obtained by taking the inverse transform of Eq. (4.71). Thus

$$x(t) = L^{-1}\frac{E(p)\,H(p)}{p} - L^{-1}\frac{G(p)\,H(p)}{p}. \tag{4.72}$$

Equation (4.72) has two terms on the right hand side which have the exact form necessary to apply the convolution (faltung) theorem of Laplace transform theory. Thus we obtain

$$x(t) = \int_0^t h(t-v)e(v)\,dv - \int_0^t h(t-v)f[x(v)]\,dv \tag{4.73}$$

which is a nonlinear integral equation for $x(t)$ of Volterra type.†

With $f \equiv 0$, the nonlinear component is absent so that

$$x^{(0)}(t) = \int_0^t h(t-v)e(v)\,dv$$

and in this notation Eq. (4.73) becomes

$$x(t) = x^{(0)}(t) - \int_0^t h(t-v)f[x(v)]\,dv. \tag{4.74}$$

---

† Initial value problems lead to Volterra type, boundary value problems lead to integral equations of Fredholm type.

At this point in the analysis we could introduce a Cauchy–Picard-like iteration which in integral equations is called the *method of successive approximations*

$$x^{(n+1)}(t) = x^{(0)}(t) - \int_0^t h(t - v) f[x^{(n)}(v)] \, dv. \qquad (4.75)$$

Uniform convergence of this sequence is covered by the below stated existence theorem of Lalesco for the nonlinear Volterra equation

$$y(t) = f(t) + \int_a^t K[t, v, y(v)] \, dv. \qquad (4.76)$$

If

(a) $f(t)$ is integrable and bounded, $|f(t)| < f$, in $a \le t \le b$;
(b) the Lipschitz condition $|f(t) - f(t')| < k|t - t'|$ holds in $a \le t \le b, k > 0$;
(c) $K(t, v, y)$ is integrable and bounded, $|K(t, v, y)| < K$ in $a \le t \le b$, $a \le v \le b, |y| < c$;
(d) the Lipschitz condition $|K(t, v, y) - K(t, v, y')| < M|y - y'|$ holds in the domain of definition of K;

then the sequence $y^{(n)}(t)$ defined via

$$y^{(0)}(t) = f(t) - f(a)$$
$$y^{(n+1)}(t) = f(t) + \int_a^t K[t, v, y^{(n)})v)] \, dv \qquad (4.77)$$

converges uniformly on $a \le t \le b$ to the solution of the integral equation (4.76).

A convenient operational procedure that has considerable utility in the calculation of the sequence $x^{(n)}(t)$ for Eq. (4.75) is based on the inverse Laplace transform of Eq. (4.69) instead of upon Eq. (4.75). From Eq. (4.69)

$$x(t) = L^{-1} y(p) = L^{-1} \frac{E(p)}{Z(p)} - L^{-1} \frac{L[f(x)]}{Z(p)}$$

whereupon the sequence $x^{(n)}(t)$ is defined by

$$x^{(0)}(t) = L^{-1} \frac{E(p)}{Z(p)},$$
$$\vdots$$
$$x^{(n+1)}(t) = x^{(0)}(t) - L^{-1} \frac{Lf[x^{(n)}(t)]}{Z(p)}. \qquad (4.78)$$

Formally, of course, the sequence $\{x^{(n)}\}$ obtained from Eqs. (4.78) is the same as that obtained from Eq. (4.75). Practically the sequence is more easily computed from Eqs. (4.78) because a table of Laplace transforms markedly assists in the analysis.

As an illustration of the foregoing procedure consider the equation

$$(a + 3cx^2)\frac{dx}{dt} + bx = e \qquad (4.79)$$

with $x(0) = 0$. Equation (4.79) has the form of Eq. (4.66) if it is rewritten in the form

$$a\frac{dx}{dt} + bx + c\frac{d}{dt}(x^3) = e. \qquad (4.80)$$

In this case we have

$$Z(D) = aD + b, \qquad f(x) = c\frac{d}{dt}(x^3), \qquad D = \frac{d}{dt}. \qquad (4.81)$$

As a result

$$H(p) = \frac{p}{Z(p)} = \frac{p}{ap + b}, \qquad h(t) = L^{-1}H(p) = \frac{1}{a}e^{-bt/a}. \qquad (4.82)$$

The integral equation satisfied by $x(t)$ is

$$x(t) = x^{(0)}(t) - \frac{c}{a}\int_0^t \exp\left[\frac{b}{a}(v - t)\right]Dx^3(v)\,dv \qquad (4.83)$$

and the operational form is

$$x(t) = x^{(0)}(t) - cL^{-1}\frac{L(Dx^3)}{ap + b} \qquad (4.84)$$

where

$$x^{(0)}(t) = L^{-1} \frac{e}{ap + b} = \frac{e}{b}(1 - e^{-bt/a}). \qquad (4.85)$$

The next approximation for x is

$$x^{(1)}(t) = \frac{e}{b}(1 - e^{-bt/a}) - cL^{-1} \frac{LD x_0{}^3}{ap + b}. \qquad (4.86)$$

Upon carrying out the indicated operations in the second term on the right and using a table of Laplace transforms we find that

$$x^{(1)}(t) = \frac{e}{b}(1 - e^{-\tau}) - \frac{ce^3}{ab^3}\left(3\tau e^{-\tau} - \frac{9}{2}e^{-\tau} + 6e^{-2\tau} - \frac{3}{2}e^{-3\tau}\right)$$

with $\tau = -bt/a$.

4.5 Application of Iterative Methods

In Section 3.5 we discussed the conversion of some similar equations in boundary layer theory to integral equation form. The method of Lock [18] Eqs. (3.127) to (3.131), is basically of a different character from that of Weyl–Siekmann (see Weyl [19]), Eqs. (3.132)–(3.142). The implications of both methods are examined.

Lock (Section 3.5) [18] considers the problem of convection in Blasius-type flow with the equations

$$f''' + ff'' = 0, \qquad f(0) = f'(0) = 0, \qquad f'(\infty) = 2, \qquad (4.87a)$$
$$\phi'' + \sigma f\phi' = 0, \qquad \phi(0) = 1, \qquad \phi(\infty) = 0 \qquad (4.87b)$$

where f and ϕ are dimensionless (similar) stream function and temperature. Note that only f' is required if we wish only the velocity component $u = U_\infty f'(\eta)$.

From Eq. (3.128) we have

$$f'' = \alpha \exp\left[-\int_0^\eta f(\eta)\, d\eta\right] \qquad (4.88)$$

and

$$f' = \alpha \int_0^\eta \exp\left[-\int_0^\eta f(\eta)\, d\eta\right] d\eta. \qquad (4.89)$$

As a first approximation, for small η, take

$$f'' = \alpha \tag{4.90}$$

whereupon $f' = \alpha\eta, f = \alpha\eta^2/2$, which we take as

$$f^{(0)} = \alpha\eta^2/2. \tag{4.91}$$

Inserting this into Eq. (4.89) we get the velocity f' as

$$f'^{(1)} = \alpha \int_0^\eta e^{-\alpha\eta^3/3}\, d\eta$$

$$= \alpha^{2/3}(3!)^{1/3} \int_0^\gamma e^{-\gamma^3}\, d\gamma$$

$$= \alpha^{2/3}(3!)^{1/3}\, \Omega(\gamma), \qquad \gamma = \left(\frac{\alpha}{3!}\right)^{1/3}\eta \tag{4.92}$$

where $\Omega(\gamma) = \int_0^\gamma e^{-\gamma^3}\, d\gamma$ is tabulated by Abramowitz [20].

The value of α is determined from the boundary condition $f'(\infty) = 2$ as

$$\alpha = \{2/(3!)^{1/3}\Omega(\infty)\}^{3/2} = 1.37 \tag{4.93}$$

(compare with the precise value of 1.3282) where $\Omega(\infty) = 0.893$ (from [20]). Thus the velocity profile is

$$u/U_\infty = f'(\eta) = (1.37)^{2/3}(3!)^{1/3}\Omega(0.611\eta)$$

$$= 2.24\,\Omega(0.611\eta). \tag{4.94}$$

Turning now to the energy Eq. (4.87b) we write it as

$$\frac{d\phi'}{\phi'} = -\sigma f\, d\eta$$

which integrates once to

$$\phi'(\eta) = C \exp\left[-\sigma \int_0^\eta f\, d\eta \right]$$

and finally to

$$\phi(\eta) = C_1 + C \int_0^\eta \exp\left[-\sigma \int_0^w f\, d\varepsilon \right] dw. \tag{4.95}$$

Setting in the boundary conditions we find $C_1 = 1$ and

$$C = C(\sigma) = -1 \Big/ \int_0^\infty \exp\left[-\sigma \int_0^w f(\varepsilon)\, d\varepsilon \right] dw.$$

Using the approximation $f'(\eta) = \alpha\eta$ we find†

$$C(\sigma) = -1 \Big/ \left[\frac{3!}{1.37\sigma} \right]^{1/3} \Gamma\!\left(\frac{4}{3}\right) e_3(\infty)$$

where $e_3(\beta) = [1/\Gamma(\tfrac{4}{3})] \int_0^\beta \exp(-\gamma^3)\, d\gamma$. From the table of Abramowitz [20] for $e_3(\infty)$

$$C(\sigma) = -0.684\, \sigma^{1/3}$$

(compare $-0.664\, \sigma^{1/3}$) and the approximate temperature profile is

$$\phi(\eta) = 1 - e_3[0.611\, \sigma^{1/3}\eta]. \tag{4.96}$$

On the other hand the Weyl–Siekmann (see Weyl [19]) approach to the Blasius problem takes an alternate route. The problem is again specified by Eq. (4.87a). Let $g = f''$ whereupon g satisfies $g' = -gf$. Integrating we find

$$g(\eta) \equiv \exp\left\{ -\int_0^\eta f(s)\, ds \right\}$$

$$= \exp\left\{ -\tfrac{1}{2} \int_0^\eta (\eta - s)^2 f''(s)\, ds \right\}$$

$$= \exp\left\{ -\tfrac{1}{2} \int_0^\eta (\eta - s)^2 g(s)\, ds \right\}, \qquad g(0) = 1 \tag{4.97}$$

(see Eq. (3.138) for the last step in this argument). This equation for g is a nonlinear integral equation for $g(\eta)$ which may be expressed in operator form by

$$g(\eta) = T\{g\} \tag{4.98}$$

where T has the following properties:

$$T(g) \geq 0 \tag{4.99a}$$

$$T(g) \geq T(g^*) \qquad \text{if} \quad g \leq g^*. \tag{4.99b}$$

---

† $\Gamma(p)$ represents the tabulated gamma function.

The first of these is obvious. If $g \leq g^*$ at all points u, $0 < u < \eta$ then

$$\int_0^\eta (\eta - s)^2 g \, ds \leq \int_0^\eta (\eta - s)^2 \, g^* \, ds$$

since g and g^* are always positive or zero. Therefore $T(g) \geq T(g^*)$. The converse is proved in the same way.

Now define the iterative sequence by

$$g^{(n+1)} = T g^{(n)} \tag{4.100}$$

and suppose we start with $g^{(0)} = 0$. In analogy with the existence theorem of Section 4.3 it follows from the inequality, Eq. (4.99b), that if $g^{(n)} \geq g^{(n+1)}$ then $g^{(n+1)} \leq g^{(n+2)}$. Thus beginning with the obvious relations $g^{(1)} > g^{(0)} = 0$ and $g^{(2)} > g^{(0)} = 0$ we develop the two sequences of inequalities

$$g^{(0)} < g^{(1)}, \quad g^{(1)} > g^{(2)}, \quad g^{(2)} < g^{(3)}, \quad g^{(3)} > g^{(4)}, \ldots \tag{4.101}$$

$$g^{(0)} < g^{(2)}, \quad g^{(1)} > g^{(3)}, \quad g^{(2)} < g^{(4)}, \quad g^{(3)} > g^{(5)}, \ldots \tag{4.102}$$

indicating oscillatory convergence. All the odd iterants are greater than the even; the odd ones decrease and the even $g^{(n)}$ increase with increasing n. Since both are monotone, bounded below and above they both converge. It is a relatively easy task to show that they both converge to the same function we call g.

The first few functions in the $g^{(n)}$ sequence are

$$g^{(0)} = 0,$$

$$g^{(1)} = 1,$$

$$g^{(2)} = \exp(-\eta^3/3),$$

$$g^{(3)} = \exp\left\{ -\int_0^\eta (\eta - s)^2 \exp(-s^3/3) \, ds \right\}.$$

In the above investigation Weyl chose $g^{(0)} = 0$. Let us express this choice in terms of the original physical problem. From Eq. (4.87a) we write

$$w''(\eta) = w''(0) \exp\left\{ -\int_0^\eta w(s) \, ds \right\}$$

so that the iteration sequence is

$$w''^{(n+1)} = w''(0) \exp\left\{ -\int_0^\eta w^{(n)}(s)\, ds \right\}. \qquad (4.103)$$

Equation (4.103) is equivalent to the differential equation

$$w'''^{(n+1)} + w^{(n)}w''^{(n+1)} = 0. \qquad (4.104)$$

The choice of $g^{(0)} = 0$ is equivalent to the initial choice

$$w^{(0)}(\eta) = \tfrac{1}{2} w''(0)\eta^2$$

as in the Lock modification.

The choice of an initial function satisfying one of the boundary conditions seems the best course of action.

More complicated examples in simultaneous nonlinear equations have been investigated by Goody (see Davies and James [2]), Davies [21], and James [22]. The latter papers concern magnetohydrodynamic boundary layer flows in various configurations.

4.6 Regular Peturbation

The method of regular perturbation should properly be called the *"small parameter method"* of Poincaré [15]. It is one of the outstanding approximate methods because it can be justified rigorously. That is to say one can establish the existence of the solution and the convergence of the series expansion in the small parameter under rather general conditions (see, e.g., Davies and James [2]). We discuss their result in terms of the pair of equations

$$\frac{dx}{dt} = f(x, y, t; \mu), \qquad \frac{dy}{dt} = g(x, y, t; \mu) \qquad (4.105)$$

where μ is a parameter.

If f and g possess a Taylor series in x, y, and μ for all values of t in $0 \le t \le t_1$ then the pair of Eqs. (4.105) have solutions for x and y which can be expressed in the form of Taylor series in the parameter μ. These series will converge when μ is sufficiently small.†

---

† This theorem, while not directly useful in the construction, does provide the analyst with the knowledge he is on the right track. "Sufficiently small" must be defined in terms of the problem at hand.

Of course there is no reason why only one "small" parameter must be used. Nowinski and Ismail [23] have utilized a multiparameter expansion in elastostatics on some linear problems. Ames and Sontowski [24] have applied multiparameter procedures to solve algebraic problems.

Extensive use has been made of the regular perturbation method in nonlinear mechanics. Some of the pertinent references include Stoker [25], McLachlan [26], Cunningham [27], and Kryloff and Bogoliuboff [28]. Bellman [29] has written the only compendium of many of the scattered results in the area and includes many up to date references.

The description of the general procedure will be given in terms of the second order equation

$$x'' + \omega^2 x = \mu f(x, x') \qquad (4.106)$$

for small μ. Equation (4.106) is equivalent to the pair of equations

$$\frac{dx}{dt} = y, \qquad \frac{dy}{dt} = \mu f(x, y) - x. \qquad (4.107)$$

Note that the *highest order derivative* is free of the parameter μ. When $\mu = 0$, $x'' + \omega^2 x = 0$ has the solution

$$x_0(t) = A \cos \omega t + B \sin \omega t \qquad (4.108)$$

which we have designated $x_0(t)$. We now assume

$$x = x_0(t) + \mu x_1(t) + \mu^2 x_2(t) + \cdots \qquad (4.109)$$

which is a *regular perturbation expansion*. For the sake of exposition assume that† $f(x, x') = x^2 + (x')^2$. With this value of f we find that substitution of Eq. (4.108) into Eq. (4.106) results in

$$(x_0'' + \mu x_1'' + \mu^2 x_2'' + \cdots) + \omega^2 (x_0 + \mu x_1 + \mu^2 x_2 + \cdots)$$
$$= \mu[x_0^2 + 2\mu x_0 x_1 + \mu^2(x_1^2 + 2x_0 x_2) + \cdots$$
$$+ x_0'^2 + 2\mu x_0' x_1' + \mu^2(x_1'^2 + 2x_0' x_2') + \cdots]. \qquad (4.110)$$

Now equating like coefficients of each power of μ (or taking successive derivatives with respect to μ and then setting $\mu = 0$) we find that the

---

† Recall that the only requirement of the theorem is that $f(x, x')$ have a Taylor series in x and x'. This is trivially a Taylor series.

original Eq. (4.106) has been transformed to the *infinite* system

$$x_0'' + \omega^2 x_0 = 0$$
$$x_1'' + \omega^2 x_1 = x_0{}^2 + x_0'^2$$
$$x_2'' + \omega^2 x_2 = 2x_0 x_1 + 2x_0' x_1'$$
$$x_3'' + \omega^2 x_3 = x_1{}^2 + 2x_0 x_2 + x_1'^2 + 2x_0' x_2'$$
$$\vdots$$

(4.111)

Let us further suppose that the initial conditions for Eq. (4.106) are

$$x(0) = \alpha, \qquad x'(0) = \beta. \tag{4.112}$$

We wish the regular expansion equation (4.109) to satisfy these initial conditions. The usual method of accomplishing this is to take

$$x_0(0) = \alpha, \quad x_0'(0) = \beta, \quad x_i(0) = 0, \quad x_i'(0) = 0, \quad i > 0. \tag{4.113}$$

The infinite system, Eq. (4.111) and Eq. (4.113), can be solved iteratively since the determination of x_k involves knowledge of x_n, $0 \leq n \leq k - 1$ in general. In developing these solutions care must be taken to avoid conditions of a resonant character—warning of such a condition is made when so-called *secular terms* appear. Such terms become unbounded as $t \to \infty$.† Typical examples include te^t, $t \cos t$, and the like.

The classic example of "secularity" occurs in the perturbation solution of the hard spring (Duffing) equation

$$x(0) = a,$$
$$x'(0) = 0, \tag{4.114}$$
$$x'' + x + \mu x^3 = 0, \qquad \mu > 0.$$

Surprises are in store here. Blithely applied regular perturbation techniques may lead to unbounded terms. Upon introducing the expansion equation (4.109) into Eq. (4.114) we find the zero order solution to be

$$x_0 = a \cos t \tag{4.115}$$

---

† Note also that such terms do not lead to periodic solutions.

and the equation for u_1 to be

$$x_1'' + x_1 = -a^3 \cos^3 t = -\frac{a^3}{4}[\cos 3t + 3\cos t],$$

$$x_1(0) = x_1'(0) = 0.$$

(4.116)

The forcing term $(a^3/4)\cos 3t$ generates a well behaved particular solution $a^3 \cos 3t/32$. On the other hand the forcing term $- (3a^3/4)\cos t$ produces the particular solution

$$-\frac{a^3}{32}(\cos t + 12t \sin t)$$

(4.117)

since $\cos t$ is a solution of the homogeneous differential equation. Thus

$$x_1 = -\frac{a^3}{32}(\cos t + 12t \sin t) + \frac{a^3}{32}\cos 3t$$

is not only *not* periodic but also becomes *unbounded* as $t \to \infty$. The term $t \sin t$ is the secular term.

One therefore observes that naive application of the regular perturbation expansion may lead to absurd results. Where is the difficulty? The answer lies in the observation that the nonlinearity affects not only the amplitude of the solution but the *frequency* as well. The difficulty is avoided if we search for the true fundamental frequency at the same time as we search for the periodic solution.

Following Poincaré and Linsted we replace the expansion $x(t) = x_0(t) + \mu x_1(t) + \mu^2 x_2(t) + \cdots$ by

$$x(t) = u_0(\omega t) + \mu u_1(\omega t) + \mu^2 u_2(\omega t) + \cdots$$

(4.118)

where each of the $u_0, u_1, \ldots$ is to be a periodic function of $\tau = \omega t$ of period 2π. The quantity ω is introduced as the true fundamental frequency which is assumed to have an expansion of the form

$$\omega = 1 + \mu\omega_1 + \mu^2\omega_2 + \cdots.$$

(4.119)

The first term is one since the original equation has unity as its fundamental frequency when $\mu = 0$.

In terms of the new variable $\tau = \omega t$ Eq. (4.114) becomes

$$\omega^2 \frac{d^2x}{d\tau^2} + x + \mu x^3 = 0.$$

(4.120)

Now, upon substituting the expansions Eq. (4.118) and Eq. (4.119) into Eq. (4.120) we obtain the sequence

$$\frac{d^2 u_0}{d\tau^2} + u_0 = 0$$

$$\frac{d^2 u_1}{d\tau^2} + u_1 = -u_0{}^3 + 2\omega_1 u_0 \qquad (4.121)$$

$$\frac{d^2 u_2}{d\tau^2} + u_2 = -3u_0{}^2 u_1 + (\omega_1{}^2 + 2\omega_2)u_0 - 2\omega_1 \frac{d^2 u_1}{d\tau^2}$$
$$\vdots$$

Clearly $u_0 = a \cos \tau$ and so the equation for u_1 becomes

$$\frac{d^2 u_1}{d\tau^2} + u_1 = 2\omega_1 a \cos \tau - a^3 \cos^3 \tau$$

$$= \left(2\omega_1 a - \frac{3a^3}{4}\right) \cos \tau - \frac{a^3}{4} \cos 3\tau \qquad (4.122)$$

where we have again used an identity for $\cos^3 \tau$ (see Eq. (4.116)). Unless the coefficient of $\cos \tau$ is zero every solution of Eq. (4.122) will contain a secular term. Thus if we take

$$\omega_1 = \frac{3a^2}{8} \qquad (4.123)$$

Eq. (4.122) becomes

$$\frac{d^2 u_1}{d\tau^2} + u_1 = -\frac{a^3}{4} \cos 3\tau$$

whose solution is $u_1 = (-a^3/32)(\cos \tau - \cos 3\tau)$.

Thus to order μ^2 we have

$$x(t) = a \cos \tau - \mu \frac{a^3}{32}(\cos \tau - \cos 3\tau) + O(\mu^2),$$

$$\omega = 1 + \mu \frac{3a^2}{8} + O(\mu^2), \qquad (4.124)$$

$$\tau = \omega t.$$

The process can be continued in like fashion using this technique (Linsted's procedure) of "casting out" the secular terms.

There are no *a priori* reasons why the zero order approximation should be linear. If a solvable zero order equation is available it is usually profitable to use it. In this direction we mention the work of Bautin [30] as modified by Davies and James [2, p. 133]. As an example we consider the boundary layer flow past a wedge where the flow outside the boundary layer is

$$U(x) = u_1 x^m. \tag{4.125}$$

From Schlichting [31] the similar equation is

$$f''' + ff'' + \beta(1 - f'^2) = 0 \tag{4.126}$$

with $f(0) = f'(0) = f'(\infty) - 1 = 0$. β is related to m through the relation

$$\beta = 2m/(m + 1).$$

For $m \ll 1$† it follows that $\beta \ll 1$. Thus we suppose f can be expanded in the regular perturbation series in β

$$f = f_0 + \beta f_1 + \beta^2 f_2 + \cdots. \tag{4.127}$$

Then the sequence $\{f_n\}$ satisfies the equations

$$f_0''' + f_0 f_0'' = 0 \tag{4.128}$$

$$f_1''' + f_0 f_1'' + f_0'' f_1 = (f_0'^2 - 1) \tag{4.129}$$

together with $f_0(0) = f_0'(0) = 0$, $f_0'(\infty) = 1$, and $f_i(0) = f_i'(0) = 0$, $f_i'(\infty) = 0$ all $i > 0$. The nonlinear problem for f_0 is essentially that of the Blasius problem (see Schlichting [31])—by a change of $\sqrt{2}$ in the Blasius problem we have this problem. The remaining equations are linear. This technique of expansion about nonlinear zero order equations has been little used.

4.7 Shohat's Expansion

Various devices have been introduced to extend the range of validity of regular perturbation expansions. One of these due to Shohat [32]† has been further amplified and applied by Bellman [33, 34]. This useful device yields results that appear to be accurate for all $\mu > 0$, not just

† We use this notation to mean m is much smaller than unity.

those for small μ. We illustrate the idea with the Van der Pol equation

$$x'' + \mu(x^2 - 1)x' + x(t) = 0, \qquad x(0) = 2, \qquad x'(0) = 0. \quad (4.130)$$

Upon setting†

$$t = s/f(\mu), \qquad f(\mu) = 1 + a_1\mu + a_2\mu^2 + \cdots \quad (4.131)$$

Eq. (4.130) transforms to

$$f^2(\mu)\frac{d^2x}{ds^2} + f(\mu)\mu(x^2 - 1)\frac{dx}{ds} + x(s) = 0. \quad (4.132)$$

If we proceed in the manner of Section 4.6 we obtain the regular expansion valid for small μ

Instead of carrying out that expansion let us introduce a new parameter r defined via ‡

$$r = \mu/(\mu + 1). \quad (4.133)$$

A variety of such transformations are probably possible. In view of what follows it is important that the transformation selected is easily manipulated. Note that for all $\mu \geq 0, 0 \leq r < 1$. From Eq. (4.133) we find

$$\mu = \frac{r}{1 - r} = r + r^2 + r^3 + \cdots,$$

$$\mu f(\mu) = r + c_1 r^2 + c_2 r^3 + \cdots. \quad (4.134)$$

We now expand $x(s)$, Eq. (4.132), in a regular perturbation expansion in r

$$x(s) = x_0(s) + rx_1(s) + r^2 x_2(s) + \cdots. \quad (4.135)$$

Substituting Eqs. (4.134) and (4.135) into Eq. (4.132) we find

$$x(s) = 2\cos s + r\sin^3 s$$
$$+ r^2\left[-\tfrac{1}{8}\cos s + \tfrac{3}{16}\cos 3s - \tfrac{5}{16}\cos 5s + \sin^3 s\right] + \cdots$$

and

$$\mu f(\mu) = r + r^2 + \frac{15}{16}r^3 + \frac{13}{16}r^4 + \cdots,$$

---

† This is related to the "coordinate stretching" method to be subsequently discussed.
‡ An Euler transformation.

that is,

$$f(\mu) = \frac{1}{1 + \mu} + \frac{\mu}{(1 + \mu)^2} + \frac{15}{16} \frac{\mu^2}{(1 + \mu)^3} + \frac{13}{16} \frac{\mu^3}{(1 + \mu)^4} + \cdots \qquad (4.136)$$

indicating a frequency correction. To illustrate the excellent accuracy of what is only supposed to be a perturbation expansion we present in Table 4.3 the values from this (4 term) expansion, Eq. (4.136), compared to Van der Pols's calculation [34a] using a numerical integration technique. A third calculation due to Urabe [35] is listed as a further check on the accuracy.

TABLE 4.3

μ	Shohat	Van der Pol	Urabe
0.1	0.998	1.00	—
0.33	0.98	0.99	—
1.00	0.93	0.90	0.92
2.00	0.77	0.78	0.81
5.60	0.43	0.38(?)	—
8.00	0.35	0.39	0.39
10.00	0.30	0.31	0.33
99.00	0.037	—	—

All the evidence points to convergence of the series (Eq. (4.136)) for all $\mu \geq 0$ but proof of this result is still lacking.

4.8 Singular Perturbation

Except for the modification of Shohat the preceding perturbation problems have involved equations of the form

$$x'' + \alpha x' + \beta x = \mu f(x, x') \qquad (4.137)$$

where f may be linear or nonlinear. Questions of greater complexity arise when the equation has the form

$$\mu x^{(n)}(t) + F[x, x', x'', \cdots, x^{(n-1)}, \mu] = 0. \qquad (4.138)$$

If a regular expansion is attempted we expect difficulties because the zero order equation, obtained by setting $\mu = 0$, is of *lower order* than the original Eq. (4.138).

To illustrate the difficulty consider the model linear problem

$$\mu f'' + f' = b, \quad f(0) = 0, \quad f(1) = 1 \qquad (4.139)$$

discussed by Friedrichs [36]. This simple model illustrates the *loss of the highest order derivative in boundary layer theory*. The exact solution to Eq. (4.139) is

$$f(x;\mu) = (1 - b)[1 - \exp(-x/\mu)]/[1 - \exp(-1/\mu)] + bx.$$

Upon attempting a regular perturbation expansion

$$f = f_0 + \mu f_1 + \mu^2 f_2 + \cdots \qquad (4.140)$$

we have $f_0' = b, f_0 = bx + c$. Thus both boundary conditions cannot be satisfied unless $b = 1$, an unlikely circumstance. The exact solution suggests that we drop the boundary condition at $x = 0$, thereby obtaining the approximation

$$f_0(x;\mu) = (1 - b) + bx. \qquad (4.141)$$

As illustrated in Fig. 4-3 this is a good approximation except within the

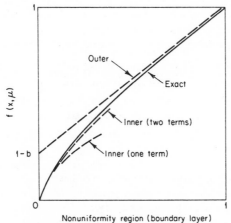

Fig. 4-3. The inner and outer solutions.

boundary layer†—that is where $x = 0(\mu)$.‡ This solution, Eq. (4.141), which is a good approximation *outside the* boundary layer is the first term of the *outer solution* (in this simple case it is the complete outer solution).

What about the *inner solution* in the boundary layer? Over the years since the end of WWII (1945) a philosophy and procedure has evolved for treating these problems. The ideas underlying the "method of inner and outer expansions" or of "double asymptotic expansions" or "the method of matched asymptotic expansions" (Bretherton [37]) have been contributed to by many. It was used by Friedrichs [38, 39] in the 1950's. It was systematically developed and applied by Kaplun [40], Kaplun and Lagerstrom [41], Lagerstrom [42], Lagerstrom and Cole [43]. A treatment of the developments and many more references is given by Van Dyke [44]. Van Dyke is a leading proponent of these methods. Most of the applications to date have been in fluid mechanics.

Corresponding mathematical efforts aimed at unraveling some of the perplexing features of equations of irregular (singular) perturbation type have closely followed the foregoing developments. Much of this work on ordinary differential equations is summarized by Wasow [45]. In addition to this monograph we also mention the earlier work of Wasow [46], Erdelyi [47], Harris [48], and Levin and Levinson [49].

Returning to our problem we now set about constructing the inner and outer expansions by the "method of matched asymptotic expansions." One of the guides should be kept in mind (Van Dyke [44]—*When terms are lost or boundary conditions discarded for the outer solution they must be included in the development of the inner solution.* Since the "inner" boundary condition was abandoned for the outer solution the "outer" boundary condition must be abandoned for the inner solution. This means that an overlap domain exists between the two solutions. The inner solution must be *matched* to the outer solution in some fashion. Kaplun and Lagerstrom [41] assert that the existence of an overlap domain between the two solutions implies that the inner expansion of the outer expansion should, to appropriate (perturbation) parameter orders of magnitude, agree with the

---

† The ideas underlying singular perturbation techniques have grown up over the years in that essentially nonlinear subject, fluid mechanics. Thus the phrase for the region adjacent to $x = 0$ as the "boundary layer."

‡ $g(x) = O[h(x)]$ as $x \to 0$ if $\lim_{x \to 0} g(x)/h(x) < \infty$. $g(x) = o[h(x)]$ as $x \to 0$ if $\lim_{x \to 0} g(x)/h(x) = 0$.

outer expansion of the inner expansion. This general matching principle is usually bent to the investigator's taste—the asymptotic form found most useful by Van Dyke [44] is the *asymptotic matching principle* used herein:

The m − term inner expansion of (the n − term outer expansion) = the n term outer expansion of (the m − term inner expansion). (4.142)

Here m and n are any two positive integers—usually we choose $m = n$ or $n + 1$. It is convenient to systematize the matching process in an orderly way. Thus beginning with the first term of the outer† expansion, we proceed as shown in Fig. 4-4. The numbers in the circles represent the term number in the respective expansion.

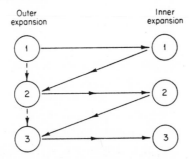

Fig. 4-4. Normal calculation sequence for the method of matched asymptotic expansions.

The first term (exact in this linear example) in the outer expansion is $f_0 = (1 - b) + bx$. To develop the first term in the inner expansion we perform a *coordinate stretching*. It is possible to stretch both dependent and independent variables or only one. Physical insight may suggest or confirm the proper stretching. Or we can seek the proper stretch by trial. The *guiding principles* dictated by the above discussion are that the inner problem must include in the first approximation any essential elements omitted in the first outer solution and that the inner and outer solutions shall match. In our problem we try stretching only the independent variable. Thus set

$$f(x; \mu) = F(X; \mu), \qquad X = x/s(\mu) \qquad (4.143)$$

---

† Generally we assign the terms outer and inner so that the outer solution is, to *first order*, independent of the inner solution.

so that Eq. (4.139) becomes

$$\left. \begin{aligned} \frac{d^2F}{dX^2} + \frac{s(\mu)}{\mu}\frac{dF}{dX} &= \frac{s^2(\mu)}{\mu}b \\ F(0) = 0, \qquad F(1/s) &= 1. \end{aligned} \right\} \qquad (4.144)$$

Recalling that the highest derivative was lost in the outer problem, d^2F/dX^2 must be kept in the inner problem. That is $s(\mu)/\mu$ must not become infinite as $\mu \to 0$. If on the other hand $s(\mu)/\mu \to 0$ as $\mu \to 0$ the resulting solution satisfying the inner boundary condition (which must also be retained) is a multiple of X. This cannot be matched to the outer solution $(1 - b) + bx$. The only remaining possibility is that $s(\mu)/\mu \to$ constant as $\mu \to 0$; this constant we may take as unity without loss of generality. Thus Eq. (4.144) become

$$\frac{d^2F}{dX^2} + \frac{dF}{dX} = b\mu, \qquad F(0) = 0, \quad F(1/\mu) = 1, \quad s(\mu) = \mu. \qquad (4.145)$$

The *inner variables* are now $X = x/\mu$ and F. Note that now the terms F'' and F' are of the same order of magnitude. Also μ now appears in the boundary condition in such a way that for $\mu \to 0$ the right hand (outer) boundary recedes to a remote position.

Let us now calculate a regular perturbation expansion $F = F_0 + \mu F_1 + \mu^2 F_2 + \cdots$ for Eq. (4.145). The zero order equation becomes

$$\frac{d^2F_0}{dX^2} + \frac{dF_0}{dX} = 0, \qquad F(0) = 0$$

whose solution is

$$F_0(X) = A(1 - e^{-X}). \qquad (4.146)$$

In determining this solution the outer boundary condition has been discarded. Imposition of the outer boundary condition would require $A = 1$ but this is incorrect as demonstrated by the exact solution—we therefore drop the outer boundary condition in the same way the inner boundary condition was dropped for the outer solution. Instead we match the inner and outer solutions using the matching principle Eq. (4.142). Our computation is arranged as follows:

1 − *term outer expansion* $= (1 - b) + bx$
    rewritten in inner variables $= (1 - b) + b\mu X$
    expanded for small $\mu = (1 - b) + \mu b X$
    1 − term inner expansion $= (1 - b)$ to $O(\mu)$
1 − *term inner expansion* $= A(1 - e^{-X})$
    rewritten in outer variables† $= A(1 - e^{-x/\mu})$‡
    expanded for small $\mu = A$
    1 − term outer expansion $= A$
    and matching shows that to first order

$$A = 1 - b. \tag{4.147}$$

To continue we must reverse the process (generally) to find the second term in the outer expansion. Thus we find $f = (1 - b) + bx +$ (no additional terms) and for the inner expansion

$$F = (1 - b)(1 - e^{-X}) + \mu[C(e^{-X} - 1) + bX].$$

Reversal of the process to find the second term in the outer expansion is not required. But if it were it would have the format $(m = 1, \text{n} = 2)$

1 − *term inner expansion* $=$
    rewritten in outer variables $=$
    expanded for small $\mu =$
    2 − term outer expansion $=$
2 − *term outer expansion* $=$
    rewritten in inner variables $=$
    expanded for small $\mu =$
    1 − term inner expansion $=$

match

We now reverse the process once more to find the second term in the inner expansion $(m = n = 2)$.

2 − *term outer expansion* $= (1 - b) + bx$
    rewritten in inner variables $= (1 - b) + \mu b X$
    expanded for small $\mu = (1 - b) + \mu b X$

---

† The outer variables are x and f.

‡ Note that $\exp[-1/\varepsilon]$ is transcendentally small compared with positive powers of ε, for any $m > 0$, $\exp[-1/\varepsilon] = o(\varepsilon^m)$.

$2-$ term inner expansion $= (1 - b) + \mu b X$

rewritten in outer variables $= (1 - b) + bx$

$2 -$ *term inner expansion* $= (1 - b)(1 - e^{-X}) + \mu[Ce^{-X} - 1) + bX]$

rewritten in outer variables $= (1 - b)(1 - e^{-x/\mu})$
$$+ \mu[C(e^{-x/\mu} - 1) + bx/\mu]$$

expanded for small $\mu = (1 - b) - \mu C + bx + \cdots$

2 term outer expansion $= (1 - b) + bx - \mu C$

rewritten in inner variables $= (1 - b) + \mu bX - \mu C.$

The additional (fifth) step in the expansions immediately above are necessary because the comparison of the two results must be made in terms of either outer or inner variables alone.

Matching in the outer variables we find $C = 0$. Thus we have the results

exact outer expansion: $f = (1 - b) + bx$

two term inner expansion: $F = (1 - a)(1 - e^{-X}) + \mu bX$

which completes the expansions to second order. The process can be continued in a similar manner.

Generally matching must proceed step by step as indicated in Fig. 4-4 (solid arrows.) Sometimes (as in the above example) one can shortcircuit the standard procedure. In an initial value problem all boundary conditions can be imposed on the outer solution. One may then compute an unlimited number of terms in the outer expansion as indicated by the dotted arrows in Fig. 4-4. These can be matched with the inner solutions to complete the calculation.

The reader should be aware that the warning provided by loss of the highest order derivative may be absent from a singular perturbation problem. The singular nature can arise from the boundary conditions and not from the equation. Additionally the singular nature can also be associated with an infinite domain. These problems are discussed in the framework of fluid mechanics by Van Dyke [44].

In any realistic problem, the coefficients in an asymptotic expansion will depend upon a variable (s), usually space or time, other than μ. The asymptotic series is said to be *uniformly valid* (in space or time) if the error is small uniformly in those variables. Examples of nonuniformity in x are

$\mu x^{-3/2} = O(\mu)$ but not uniformly near $x = 0;$

$\mu \log x = O(\mu)$ but not uniformly near $x = 0, \infty.$

Van Dyke [44] gives the following definition of a singular perturbation problem: *A singular perturbation problem is one in which no single asymptotic expansion is uniformly valid throughout the field of interest.*

4.9 The Method of Strained Coordinates

As an outgrowth of research involving bow shock waves in supersonic flow Lighthill [50] described a general technique for removing non-uniformities from perturbation solutions of nonlinear problems. This method of Lighthill† or *method of strained coordinates* constitutes a useful alternative to the method of matched asymptotic expansions.

Lighthill observed that the (zero order) linearized solution may have the right form, but not quite at the right place. To cure this defect he suggested slightly straining the coordinates, by expanding one of them as well as the dependent variable in asymptotic series (compare with the method of Shohat). The first approximation is therefore rendered uniformly valid. The coordinate straining is initially unknown and must be determined term by term as the solution progresses.

The straining is determined by Lighthill's principle:

Higher approximations shall be no more singular than the first. (4.148)

The development of strained coordinates using this principle halts the compounding of singularities that invalidates a straightforward regular expansion in a region of nonuniformity. Uniform validity is restored in a wide class of problems.

Lighthill's principle is not a definite procedure but rather represents a philosophy. Thus the principle does not determine the straining *uniquely*. This nonuniqueness can often be used to great advantage. Since both dependent and independent variables are expanded, the solution is obtained parametrically, with the strained coordinate appearing as a parameter.

The method has proved useful for hyperbolic problems—particularily for those with unidirectional travelling waves. Lin [51] has generalized the method so that it applies to hyperbolic equations in two variables. Lighthill [52] advises that the method be used for hyperbolic systems only.

---

† Van Dyke [44] discusses the historical background of the method. Poincaré [15] (1892) also strained the independent variable in a similar way. Van Dyke's main concern is with partial differential equations.

Lighthill [50] introduces his method by considering the general equation

$$(x + \mu y)\frac{dy}{dx} + q(x)y = r(x). \tag{4.149}$$

We illustrate the concepts with the equation

$$(x + \mu y)\frac{dy}{dx} + y = 1, \qquad y(1) = 2 \tag{4.150}$$

whose exact solution is

$$y = \left[\left(\frac{x}{\mu}\right)^2 + 2\frac{1 + x}{\mu} + 4\right]^{1/2} - x/\mu, \tag{4.151}$$

which is *finite* at $x = 0$ with the value

$$y(0) = [4 + 2/\mu]^{1/2}.$$

When a regular perturbation expansion in μ is carried out one obtains the series

$$y = \frac{1 + x}{x} - \mu\frac{(1 + 3x)(1 - x)}{2x^3} + \mu^2\frac{(1 + x)(1 - x)(1 + 3x)}{2x^5} + \cdots. \tag{4.152}$$

The series converges but as $x \to 0$ the radius of convergence vanishes. The expansion is therefore not uniformly valid near $x = 0$. The original equation has a singularity along the line $x + \mu y = 0$. Linearization by perturbation transfers the singularity to $x = 0$ (see Eq. (4.152)) and this singularity is intensified to $1/x^3$ and $1/x^5$ in higher approximations rather than being corrected. A picture of the true situation and what happens in the perturbation is shown in Fig. 4-5 as adapted from the work of Van Dyke [44].

Lighthill's method allows the singularity in the first approximation to shift toward its true position by straining the independent coordinate x. The straining is accomplished by expanding both y and x in powers of μ

$$x = s + \mu x_1(s) + \mu^2 x_2(s) + \cdots \tag{4.153}$$

$$y = y_0(s) + \mu y_1(s) + \mu^2 y_2(s) + \cdots \tag{4.154}$$

with coefficient functions depending upon the new auxiliary variable $x_0(s) = s$. $x_0(s)$ may be chosen as s since the straining is slight.

When Eqs. (4.153) and (4.154) are substituted into Eq. (4.150) and like powers of μ are equated we find

$$(sy_0)' = 1 \tag{4.155a}$$

$$(sy_1)' = y_0'(sx_1' - x_1 - y_0)$$
$$= [x_1(1 - y_0) - \tfrac{1}{2}y_0^2]' \tag{4.155b}$$
$$\vdots$$

The solution for y_0, such that $y_0(1) = 2$ is

$$y_0(s) = (1 + s)/s. \tag{4.156}$$

Setting Eq. (4.156) into Eq. (4.155b) we find the equation for y_1 to be

$$(sy_1)' = -\left[\frac{x_1}{s} + \frac{(1 + s)^2}{2s^2}\right]'$$

which integrates to

$$y_1(s) = \frac{C}{s} - \frac{1}{s}\left[\frac{x_1(s)}{s} + \frac{(1 + s)^2}{2s^2}\right]. \tag{4.157}$$

The straining $x_1(s)$ is to be chosen in accord with Lighthill's principle, Eq. (4.148). Here this means that y_1 shall be no more singular than y_0. There are many ways of accomplishing this. One way is to choose $x_1(s)$ so that

$$\frac{x_1(s)}{s} + \frac{(1 + s)^2}{2s^2} = \text{const} = C_1 \tag{4.158}$$

but let us also note that the constant can be replaced by any function $g(s)$ having a Taylor series. Let us select $C_1 = 0$ so that

$$x_1(s) = -(1 + s)^2/2s. \tag{4.159}$$

A uniform first approximation, in parametric form, is therefore

$$y \approx \frac{1 + s}{s} + \cdots$$

$$x \approx s - \mu\frac{(1 + s)^2}{2s} + \cdots. \tag{4.160}$$

In this simple problem one can eliminate the parameter s and obtain

$$y = \left(\left(\frac{x}{\mu}\right)^2 + \frac{2(1 + x)}{\mu}\right)^{1/2} - \frac{x}{\mu}$$

whose value at $x = 0$ is $[2/\mu]^{1/2}$ which agrees with the exact solution to first order in μ (see Fig. 4-5).

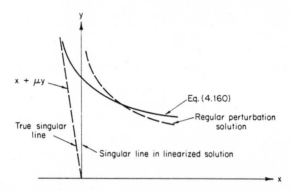

Fig. 4-5. Illustration of the method of strained coordinates.

A second way to choose the straining is to make it vanish at the position $x = 1$ where the boundary condition is applied. To accomplish this we note that $x = 1$ corresponds to $s = 1$ so that $x_n(1) = 0$ for all $n > 0$. In Eq. (4.157) let us first evaluate C, noting that $y_1(1) = 0$. Thus

$$0 = C - [0 + 2], \qquad C = 2$$

and

$$y_1(s) = \frac{2}{s} - \frac{1}{s}\left[\frac{x_1(s)}{s} + \frac{(1 + s)^2}{2s^2}\right]. \tag{4.161}$$

To make the straining vanish we need only select

$$x_1(s) = (3s^2 - 1 - 2s)/2s \tag{4.162}$$

whereupon the uniformly valid first approximation becomes

$$y \approx \frac{1 + s}{s} + \cdots, \qquad x \approx s + \mu\left(\frac{3s^2 - 1 - 2s}{2s}\right). \tag{4.163}$$

4.10 Comparison of the Two Preceding Methods

The reader may find it instructive to develop the inner and outer solution by the method of matched asymptotic expansions for the problem of Section 4.9. For comparison we shall include only the barest details. We use x and y as the outer variables and Eq. (4.152) as the outer expansion. In the inner variables $X = x\mu^{-1/2}$, $F = y\mu^{1/2}$ Eq. (4.150) becomes

$$(X + F)\frac{dF}{dX} + F = \mu^{1/2}. \tag{4.164}$$

The outer expansion equation (4.152) proceeds in integral powers of μ, but the inner expansion proceeds by half powers as

$$F = F_0 + \mu^{1/2} F_1 + \cdots. \tag{4.165}$$

By the method of matched expansions we find the first order approximations to be (Van Dyke [44])

$$y \approx \begin{cases} \dfrac{1 + x}{x} & \text{(from outer)}, \quad \mu \to 0 \quad \text{with} \quad x \quad \text{fixed} \\[2ex] \left[\left(\dfrac{x}{\mu}\right)^2 + \dfrac{2}{\mu}\right]^{1/2} - \dfrac{x}{\mu} & \text{(from inner)}, \quad \mu \to 0 \quad \text{with} \quad x\mu^{-1/2} \quad \text{fixed}. \end{cases} \tag{4.166}$$

In this example one sees that Lighthill's method is by far the simpler. A strength of Lighthill's method, as well as a weakness, is that only one asymptotic expansion is used for the dependent variable. This usually results in a simpler analysis that often avoids an intractable nonlinear problem for the first term of the inner expansion. However, the inexact nature of the straining principle compared with the detailed rules of the matching process work against the method of strained coordinates. Van Dyke [44] observes that the method of matched asymptotic expansions is the more reliable and perhaps the more fundamental of the two methods.

Lighthill's method may appear to succeed while giving incorrect results. It cannot be used in a problem that is singular because the highest derivative is multiplied by the small parameter. Further discussion of these points may be found in the work of Van Dyke [44].

The convergence of Lighthill's method when applied to Eq. (4.149) has been examined by Lighthill [52] and Wasow [53]. Modifications by Temple have *regularized* the method as discussed by Davies and

James [2]. Sandri [54–56] has developed a general method for uniformizing that includes the Poincaré and Lighthill procedure. Bowen *et al.* [57] obtain a singular perturbation refinement to quasisteady state approximations in chemical kinetics. Their tool is the inner-outer expansion of the matching technique applied to both linear and nonlinear equations.

Free [58] (see Davies and James [2, p. 155]) applies the inner-outer solution method to a hydrodynamical problem. Carrier and Lewis [59] investigate Van der Pol's equation

$$\frac{d^2x}{dt^2} + \mu(x^2 - 1)\frac{dx}{dt} + x = 0$$

for large μ. If we set $t = \mu\tau$, $\varepsilon = \mu^{-2}$, this equation transforms to

$$\varepsilon\frac{d^2x}{d\tau^2} + (x^2 - 1)\frac{dx}{d\tau} + x = 0$$

where ε is a small parameter. It is clear that this is singular.

4.11 Weighted Residual Methods—General Discussion

A desire for direct methods in the calculus of variations led Rayleigh (for eigenvalue problems) and Ritz (for equilibrium problems) to develop powerful methods which have been widely used in linear problems. Later Galerkin [60] in 1915 developed the first true weighted residual method. The background for these methods and examples of their applications to linear problems are available in many sources—here we content ourselves with listing three, Kantorovich and Krylov [61], Collatz [62], and Crandall [63]. In this section we shall discuss the general method and then detail various error distribution methods.

Let x be the independent variable and suppose the problem is formulated in a domain D as

$$\begin{aligned} L[u] &= f(x) \\ B_i[u] &= g_i(x), \qquad i = 1, 2, ..., p \end{aligned} \tag{4.167}$$

where L is a nonlinear differential operator (of course it could be linear), B_i represent the appropriate number of boundary conditions and f, g_i are functions.

We seek an approximate solution to Eq. (4.167) in the linear form

$$\bar{u}(x) = \sum_{j=1}^{n} C_j \phi_j + \phi_0 \qquad (4.168)$$

where the $\phi_j, j = 1, 2, \ldots, n$ is a set of "trial" functions chosen before-hand. The ϕ_j are *often* chosen to satisfy the boundary conditions, indeed the original Galerkin method was of this type. This requirement may be modified and its modification will be discussed subsequently. We can always consider the chosen functions to be linearly independent and to represent the first n functions of some set of functions $\{\phi_i\}$, $i = 1, 2, \ldots$ which is complete† in the given domain of integration. The ϕ_j are functions of the independent variable so that the C_j are *undetermined parameters*. There are two basic types of criteria for fixing the C_j. In the *weighted residual method* the C_j are so chosen as to make a weighted average of the equation residual vanish. In the other (which in reality is not a weighted residual method but has so many common features that we treat it here), the C_j (as undetermined parameters) are chosen to give a stationary value to a functional related to Eq. (4.167). This functional is usually obtained via the calculus of variation. In both cases, one obtains a set of n simultaneous (nonlinear) algebraic equations for the C_j. The generalizations for partial differential equations are given by Ames [64].

In the weighted residual methods the trial solution Eq. (4.168) is *usually*‡ chosen to satisfy all the boundary conditions in both equilibrium and initial value problems. This can be accomplished in many ways. The suggested method is to choose the ϕ_j to satisfy

$$\begin{aligned} B_i[\phi_0] &= g_i, & i = 1, \ldots, p \\ B_i[\phi_j] &= 0, & i = 1, \ldots, p; \quad j \neq 0. \end{aligned} \qquad (4.169)$$

It is then clear that $\bar{u}$ satisfies all the boundary conditions. However, in the case of *initial value problems* the initial conditions often cannot also be satisfied and a separate initial residual is established.

---

† In a function space (e.g., the set of all continuous functions in $0 \leq x \leq 1$) a complete set $\{\phi_i\}$ is defined as a set such that no function $F(x)$ exists in the space that cannot be expanded in terms of the $\{\phi_i\}$.

‡ A discussion of possible generalizations is given in Section 4.13.

For the stationary functional method it is only necessary that $\bar{u}$ satisfy the *essential*† boundary conditions.

When the trial solution, Eq. (4.168) is substituted into Eq. (4.167) the equation residual R becomes

$$R[C, \phi] = f - L[\bar{u}] = f - L[\phi_0 + \sum_{j=1}^{n} C_j\phi_j(x)], \qquad (4.170)$$

where the notation $R[C, \phi]$ indicates the dependence of R on the vectors $C = (C_1, \ldots, C_n)$ and $\phi = (\phi_0, \phi_1, \ldots, \phi_n)$. When $\bar{u}$ is the exact solution R is identically zero. Within a restricted trial family a *good* approximation may be described as one in which R is small *in some sense*.

A. STATIONARY FUNCTIONAL CRITERION‡

Let Ω be a functional (e.g., an integral) derived from the original problem. This (Ritz) method consists in inserting the trial family directly into Ω and asking for

$$\frac{\partial \Omega}{\partial C_j} = 0, \qquad j = 1, 2, \ldots, n. \qquad (4.171)$$

These n equations are solved for the C_j and the corresponding function $\bar{u}$ represents an approximate solution to the extremum problem. The n algebraic equations are generally nonlinear for a nonlinear equilibrium problem.

B. WEIGHTED RESIDUAL CRITERIA*

The sense in which the residual R is small is that each of the weighted averages,

$$\int_D W_k R \, dx, \qquad k = 1, 2, \ldots, n, \qquad (4.172)$$

---

† Essential boundary conditions can be described by the Collatz [62] condition: if the differential equation is of order $2m$, then the essential boundary conditions are those that can be expressed in terms of u and its first $m - 1$ derivatives. In elasticity the essential conditions are those of geometric compatibility and the *natural ones* are those of force balance.

‡ For equilibrium problems.

* Stated for equilibrium problems. Modifications for initial value problems are discussed later in the section.

of R with respect to the weighting function $W_k, k = 1, 2, \ldots, n$ should vanish. This requirement provides n algebraic, usually nonlinear, equations for the nC_j if the trial solution, Eq. (4.168), is selected in the linear form. However, in some cases nonlinear (in the C_j) trial solutions may be chosen that will lead to linear equations. Various cases can be distinguished and *certainly this list is not complete.*

(i) *Method of Moments*

In this case $W_k = P_k(x)$, so Eq. (4.172) becomes $\int_D P_k(x) R\, dx = 0$ which is equivalent to asking for the vanishing of the first n "moments" of R. Here $P_k(x)$ are orthogonal polynomials in the vector x over the domain D. Some authors use $x^k = W_k$ but these are not orthogonal on the interval $0 \leq x \leq 1$ and better results would be obtained if they were orthogonalized before use. The use of x^k gives the method its name "method of moments."

(ii) *Collocation*

In this case we choose n points in the domain D, say $p_i, i = 1, 2, \ldots, n$, and let

$$W_k = \delta(p - p_k) \qquad (4.173)$$

where δ represents the unit impulse or Dirac delta which vanishes everywhere except at $p = p_k$ and has the property that $\int_D \delta(p - p_k) R\, dx = R(p_k)$. This criterion is thus equivalent to setting R equal to zero at n points in the domain D. The point location is arbitrary but is usually such that D is covered by a simple pattern. Special circumstances of the particular problem may dictate other patterns.

(iii) *Subdomain*

Here the domain D is subdivided into n subdomains, not necessarily disjoint, say $D_1, D_2, \ldots, D_n$. The weighting functions W_k are chosen as

$$W_k(D_k) = 1, \qquad W_k(D_j) = 0 \qquad j \neq k, \qquad (4.174)$$

so that the closeness criterion, Eq. (4.172), becomes

$$\int_{D_k} R\, dx = 0, \qquad k = 1, 2, \ldots, n.$$

(iv) *Least Squares*

The integral of the square of the residual is minimized with respect to the undetermined parameters to provide the n simultaneous equations for the C_j. Then

$$\frac{\partial}{\partial C_k} \int_D R^2 \, dx = 2 \int_D \frac{\partial R}{\partial C_k} R \, dx = 0, \quad k = 1, 2, ..., n \qquad (4.175)$$

so we infer that $W_k = \partial R / \partial C_k$.

(v) *Galerkin's Method*

Here we choose $W_k = \phi_k$, where the ϕ_k are the portion of the complete (and hopefully orthogonal) set used to construct the trial solution $\bar{u}$. Thus Galerkin's method asks for

$$\int_D \phi_k R \, dx = 0, \quad k = 1, 2, ..., n. \qquad (4.176)$$

When the problem is an initial value (propagation) problem the trial function should be so selected that the initial conditions are satisfied. Since the range of the time variable is infinite an estimate of the steady state (if any) and when it is approximately achieved will be helpful in establishing what time interval need be considered. If the steady solution is obtainable it may be used as the asymptotic solution for $t \to \infty$ and an approximate solution can be calculated for small t by one of these weighted residual methods.

It is clear from the preceding discussion that the most important and most difficult step, in all these methods, is the selection of the trial solution, Eq. (4.168). Application of the criteria to this trial solution has the effect of selecting the "best" approximation from the given family. One must ensure that good approximations are included within the trial family. In selecting the ϕ_j one should carefully insure that the functions are (a) linearly independent; (b) members of a complete set; (c) incorporating special characteristics of the problem such as symmetry, monotonicity, periodicity, etc which may be known.

When Eq. (4.167) is *linear* the algebraic equations for the C_j's obtained by any of the weighted residual methods will be linear. Moreover, the matrix of the coefficients of the C_j will always be symmetric in the least squares method but not generally for the other methods. However if the problem, Eq. (4.167), is self-adjoint† then Galerkin's criterion will also generate symmetric equations. When the functional Ω is quadratic, the Ritz method generates symmetric linear equations for the C_j.

When these methods are applied to a particular trial family satisfying all the boundary conditions they generally produce different approximations. In the case of a *linear* equilibrium problem having an equivalent extremum formulation the Ritz method and Galerkin's method (applied to the same trial family) give identical results (see Galerkin [60]). Thus in this case Galerkin's criteria provides the optimum of the true weighted residual methods in the sense that the approximation so obtained also renders Ω stationary.

Numerous investigations have been carried out comparing these methods on the same problem. Comparisons on linear problems in partial differential equations have been made (on system properties) in equilibrium problems by Crandall [63, p. 234] and point by point with the exact solution in Hetenyi [65]. Eigenvalue comparisons have been detailed by Crandall [63, p. 318] and Frazer *et al.* [66]. Bickley [67] considered an initial value problem of heat transfer. His approximations using two undetermined functions compared collocation on both the equation and initial residual, collocation on the initial residual with moments on the equation residual, moments of both residuals and Galerkin on both residuals. All of these *linear* comparisons demonstrate the general superiority of Galerkin's method. *All of this experience does not constitute proof and the reader is cautioned that there may be examples where other methods are superior.*

Two recent reviews of this subject are excellent. In the first of these Finlayson and Scriven [68] review the literature and the present standing of the theory of the weighted residual methods. In the second Finlayson and Scriven [69] discuss the several attempts to formulate variational principles for non-self-adjoint and nonlinear systems. The corresponding variational methods of approximation are shown to be equivalent to the more straightforward Galerkin method or a closely related version of the method of

---

† When the system equation (4.167) is linear in both the equation and boundary conditions we say it is self-adjoint if for any two functions u and v satisfying the homogeneous boundary condition $B_i[u] = B_i[v] = 0$ for all i, we always have $\int_D uL[v] \, dD = \int_D vL[u] \, dD$.

weighted residuals. They conclude that there is no practical need for variational formulations of the sort examined.

After several examples we shall return to a further discussion of these methods.

4.12 Example of Weighted Residuals

In this section we describe an application of the foregoing weighted residual methods. The example selected uses a modification of the collocation process called *extremal point collocation* due to Jain [70].

The problem of finding an approximate solution of the boundary value problem Eq. (4.167) is often attacked using the trial function equation (4.168) which depends on a number of parameters $C_1, C_2, \ldots, C_n$ and is such that for arbitrary values of the C_j

(a)   the equation is satisfied exactly, or
(b)   the boundary conditions are exactly satisfied, or
(c)   neither the differential equation nor the boundary conditions are exactly satisfied.

Our goal is to determine the parameters C_j so that $\bar{u}(x) = \bar{u}(x, C_j)$ satisfies in case (a) the boundary conditions, in case (b) the differential equation, in case (c) both, as accurately as possible. The free parameters and the estimation of error are determined by suitable choice of a norm (distance function). Since a great variety of norms are available (see, e.g., Collatz [62]) there exists a correspondingly large number of methods of collocation and other methods. Some of the traditional error estimates and approximation methods are unsuitable for numerical analysis. One of these, based upon the Hilbert (L_2) norm

$$\|f\| = \left[\int_D f^2 \, dD\right]^{1/2},$$

is not very useful because $\|f\|$ can be very small while f itself can assume very large values. An alternate norm is the max-norm

$$\|f\| = \max_D |f| \tag{4.177}$$

which is much used in Banach spaces. In addition, as discussed by Collatz [71], it has acquired special usefulness in numerical analysis. We shall use this norm to control the distribution of the residual in the sequel.

Based upon Eq. (4.177) the extremal point collocation procedure requires that the residual, written $R[x, C]$, either from the equation or from the boundary conditions or both, possess alternate maximum and minimum values. For n undetermined coefficients we choose $(n + 1)$ points $x_0, x_1, \ldots, x_n$ in the interval $D : a \leq x \leq b$, and set, for each $j = 1, 2, \ldots, n$,

$$R[x_j, C] - (-1)^j R[x_0, C] = 0. \qquad (4.178)$$

This constitutes a system of n equations, usually nonlinear. Furthermore, put

$$R[x_j, C] - (-1)^i R[x_0, C] = D_i(C). \qquad (4.179)$$

Let $C_k^{(0)}$, $k = 1, 2, \ldots, n$ represent the coefficients of a trial solution and suppose the actual solution is $C_k^{(0)} + \eta_k$. Then one finds by series expansion that

$$D_i[C_1^{(0)} + \eta_1, C_2^{(0)} + \eta_2, \ldots, C_n^{(0)} + \eta_n]$$

$$= D_i[C_1^{(0)}, \ldots, C_n^{(0)}] + \sum_{j=1}^{n} \eta_j \frac{\partial D_i}{\partial C_j^{(0)}} + O(\eta^2), \qquad i = 1, 2, \ldots, n.$$
$$(4.180)$$

Let

$$Y = [Y_{ij}^{(0)}] = \begin{bmatrix} \dfrac{\partial D_1}{\partial C_1^{(0)}} \cdots \dfrac{\partial D_1}{\partial C_n^{(0)}} \\ \cdots \\ \dfrac{\partial D_n}{\partial C_1^{(0)}} \cdots \dfrac{\partial D_n}{\partial C_n^{(0)}} \end{bmatrix}. \qquad (4.181)$$

Those values of η_j for which

$$\sum_{j=1}^{n} \frac{\partial D_i}{\partial C_j^{(0)}} \eta_j = -D_i[C_1^{(0)}, \ldots, C_n^{(0)}]$$

or in matrix form

$$Y\eta = -D, \qquad D \text{ the vector of the } D_i,$$

are uniquely determined by†

$$\eta = -Y^{-1}D \qquad (4.182)$$

_____
† η and D are column vectors and Y^{-1} denotes the inverse matrix.

if the determinant of Y is not zero. These corrections correspond to an extension of the ordinary Newton–Raphson process.

Equation (4.182) in component form can be written as

$$\eta_j = -[Y_{ij}^{(0)}]^{-1}D_i[C_1^{(0)}, ..., C_n^{(0)}].$$

Thus an iterative process is easily inferred to be of the form

$$C_i^{(k+1)} = C_i^{(k)} - [Y_{ij}^{(k)}]^{-1}D_i[C_1^{(k)}, ..., C_n^{(k)}] \qquad (4.183)$$

where $[Y_{ij}^{(k)}]$ denotes the matrix analogous to Eq. (4.181) evaluated at the kth step in the iteration. The iterative process, Eq. (4.183), is second order.

If the extremal collocation points are selected as the extremal points

$$x_n^{(\mu)} = \cos\left(\frac{n\pi}{\mu}\right) \qquad (4.184)$$

of the Chebyshev polynomials, the residual error curve will be fairly uniform over the interval. The points $x_n^{(\mu)}$ can be improved by applying the iteration method

$$R[x_j^{(v)}, C] - (-1)^j R[x_0^{(v)}, C] = 0 \qquad (4.185)$$

where $x_j^{(v)}$ and $x_0^{(v)}$ represent the extremal points of $R[x, C]$ at the vth step of the computation. Usually a better and quicker result is obtained by setting

$$\frac{dR[x, C]}{dx}\bigg|_{x=x_i} = 0. \qquad (4.186)$$

Jain [70] has computed two examples from fluid mechanics by this method. In the more complicated example of flow near a rotating disc the similar equations are

$$2F + H' = 0$$
$$F^2 + HF' - G^2 - F'' = 0$$
$$2FG + HG' - G'' = 0$$
$$P' + HH' - H'' = 0$$

at

$$\xi = 0: \quad F = 0, \quad G = 1, \quad H = 0, \quad P = 0$$
$$\xi = \infty: \quad F = 0, \quad G = 0.$$

In the second problem, which we consider here, the similarity problem for Newtonian flow near a stagnation point generates the system

$$\left.\begin{array}{l} \phi''' + 2\phi\phi'' - \phi'^2 + 1 = 0 \\ \xi = 0: \quad \phi = 0, \quad \phi' = 0 \\ \xi = \infty: \quad \phi' = 1. \end{array}\right\} \tag{4.187}$$

Our trial solution will be of the form

$$f(\xi) = \phi_0(\xi) + C_1\phi_1(\xi) + C_2\phi_2(\xi) \tag{4.188}$$

where the linearly independent ϕ_j's are each chosen to satisfy *all* the boundary conditions

$$\begin{aligned} \phi_0 &= \xi - 1 + e^{-\xi} \\ \phi_1 &= 1 - 2e^{-\xi} + e^{-2\xi} \\ \phi_2 &= 2 - 3e^{-\xi} + e^{-3\xi} \end{aligned} \tag{4.189}$$

Upon setting Eq. (4.188) and Eq. (4.189) into Eq. (4.187) we find the residual

$$\begin{aligned} R[\xi, C] &= \phi_0''' + C_1\phi_1''' + C_2\phi_2''' + 2[\phi_0 + C_1\phi_1 + C_2\phi_2] \\ &\times (\phi_0'' + C_1\phi_1'' + C_2\phi_2'') - (\phi_0' + C_1\phi_1' + C_2\phi_2')^2 + 1 \end{aligned} \tag{4.190}$$

for the three term expansion.

To carry out the computation we take

$$R[\xi_j^{(v)}, C] - (-1)^j R[\xi_0^{(v)}, C] = 0, \quad j = 1, 2 \tag{4.191}$$

beginning with

$$\xi_0^{(1)} = 0, \quad \xi_1^{(1)} = 0.5, \quad \xi_2^{(1)} = 1.0$$

and iterate, at the same time computing the coefficients C_i by means of Eq. (4.183). The results are shown in Table 4.4.

The approximation is therefore

$$\begin{aligned} f(\xi) &= \xi - 1 + e^{-\xi} + 0.71722755(1 - 2e^{-\xi} + e^{-2\xi}) \\ &- 0.18758077(2 - 3e^{-\xi} + e^{-3\xi}). \end{aligned} \tag{4.192}$$

It is interesting to compare the value of $f''(0)$ with other methods:

Exact: 1.3120
Collocation: 1.309
Integral method: 1.255
Tomoita: 1.283.

TABLE 4.4

	Collocation points, residual			Extremal points	
$\zeta_j^{(1)}$	0	0.5	1.0	0.3	1.4
$R[\zeta_j^{(1)}]$	0.165069	−0.165069	0.165069	−0.254257	0.166181
$\zeta_j^{(2)}$	0	0.3	1.4	0.282835	1.41166
$R[\zeta_j^{(2)}]$	0.198296	−0.198296	0.198296	−0.199014	0.198328
$\zeta_j^{(5)}$	0	0.283702	1.41176	0.283702	1.41176
$R[\zeta_j^{(5)}]$	0.198573	−0.198573	0.198573	−0.198573	0.198573

Coefficients	$\nu = 1$	$\nu = 2$	$\nu = 5$
$C_1^{(\nu)}$	0.77719302	0.71768623	0.71722755
$C_2^{(\nu)}$	−0.20117615	−0.18768388	−0.18758077

4.13 Comments on the Method of Weighted Residuals

By this point it is clear that one should construct the trial solution so that a maximum of information can be extracted with a minimum of computation. The more we know about the expected behavior of a solution the better a trial solution can be made. In propagation problems a preliminary study of the expected behavior is very important. It is sometimes unwise to attempt to approximate the solution by a single approximate solution throughout. The solution domain should be broken up and the separate zones individually treated.

The determination of the undetermined parameters or functions in all of these methods can be thought of as utilizing certain *error distribution principles*. These principles distribute the error in the approximate solution over the domain of the problem according to a prescribed criterion. The method of moments and the Galerkin method are special cases of "orthogonality" methods which specify that the error should be orthogonal to a chosen set of linearly independent weighting functions. The error distribution methods, as opposed to the Ritz and other stationary functional methods, have the advantage that they work directly with the differential equations rather than with an equivalent variational problem.

We have pointed out that one should take the trial function members ϕ_j in order from a complete set. While any complete set may be used it is convenient to choose an expansion which identically satisfies either the boundary conditions (as we stated previously) or the differential equations. One can usually obtain an accurate approximation with few terms in such an event.

Basically there are three variations of the method;

(1) *Interior Method*

The trial functions are chosen so that the boundary conditions are identically satisfied. Since only the interior error R_E is not zero it is distributed according to one of the error distributions of Section 4.11.

(2) *Boundary Method*

The trial functions are chosen so that the differential equations are identically satisfied. The only sources of error are those of the boundary errors R_B. These errors are distributed according to one of the error distributions of Section 4.11.

(3) *Mixed Method*

In some situations it may not be feasible to choose a set of trial functions which is of type (1) or (2). For such systems all three types of error must be distributed.

Associated with any approximate procedure there is always a question regarding the accuracy of approximation. In many cases there are theorems to the effect that if an iteration is continued without limit, or if an increment size is decreased without limit, or if the number of parameters is increased indefinitely etc, then the process converges to the exact solution. Such results have value in that they encourage the analyst to adopt a procedure *but* it is seldom possible to obtain more than one or five or 200 steps (with a computer) of an infinite process. *Therefore a realistic error bound applicable at any stage of the calculation is of considerably more value than convergence theorems.* Convergence theorems are relatively easy to prove while *realistic* error bounds are relatively rare because they are difficult to obtain, even in linear systems. If no error analysis is available a common procedure for examining the power of an approximate method is to apply it to a problem whose exact solution is already known. The error can then be exactly determined and the presumption is made that in similar problems the method will produce errors of the same order of magnitude. Such arguments are far from ideal but are often the only practical way out for most nonlinear problems.

4.14 Quasilinearization

The origin of quasilinearization lies in the theory of dynamic programming. The background and applications to other areas than the one which concerns us here are given by Bellman and Kalaba [72]. According to these authors the objectives of the theory are (1) A uniform approach to the study of existence and uniqueness of solutions of (nonlinear) ordinary and partial differential equations subject to initial and boundary conditions; (2) To obtain representation theorems for these solutions in terms of the solutions of linear equations; (3) To obtain a uniform approach to the numerical solution of both descriptive and variational problems—an approach which possesses various monotonicity properties and rapidity of convergence. The numerical algorithms are meant for execution on digital computers.

To motivate the method we recall the Newton–Raphson development of a sequence of approximations to a root of the scalar equation $f(x) = 0$. We assume† $f(x)$ is monotone decreasing, that is $f'(x) < 0$ for all x and is strictly convex, that is $f''(x) > 0$. If $x_0 < r, f(x_0) > 0$ constitutes our initial approximation to the root r, then one develops by series expansion the general Newton–Raphson recurrence relation

$$x_{n+1} = x_n - \frac{f(x_n)}{f'(x_n)}. \tag{4.193}$$

From the available inequalities $f(x_n) > 0, f'(x_n) < 0$ for each n one easily demonstrates that the sequence equation (4.193) is monotone in its convergence—that is, if $x_0 < r$, then $x_0 < x_1 < x_2 < x_3 < \cdots < r$.

Even more important computationally, but not so obvious, is the

---

† The method is more general than that discussed herein.

second property of *quadratic convergence*. This property of the sequence, Eq. (4.193), in one form, asserts that

$$|x_{n+1} - r| \le k |x_n - r|^2 \qquad (4.194)$$

where k is independent of n and $k = \max_{x_0 \le \theta \le r} |\phi''(\theta)|/2$ with $\phi'(x) = f(x)f''(x)/[f'(x)]^2$. An alternate,

$$|x_{n+1} - x_n| \le k_1 |x_n - x_{n-1}|^2, \qquad (4.195)$$

where

$$k_1 = \max_{x_0 \le \theta \le r} \left[\frac{|f''(\theta)|}{|f'(\theta)|} + \frac{|\phi''(\theta)|}{2} \right]$$

is also called quadratic convergence.

As x_n approaches r there is a tremendous acceleration of convergence. Asymptotically, each additional step doubles the number of correct digits.

In previous chapters we developed the relation between the linear equation

$$u'' + p(t)u' + q(t)u = 0 \qquad (4.196)$$

and the Riccati equation

$$v' + v^2 + p(t)v + q(t) = 0. \qquad (4.197)$$

Recall that Eq. (4.196) transforms into Eq. (4.197) if we set $u = \exp[\int v \, dt]$ or $v = u'/u$. We now discuss the solution of Eq. (4.197) in terms of a *maximum operation*.

To initiate our discussion consider the parabola $y = x^2$ and the tangent at a point $(x_1, y_1) = (x_1, x_1^2)$

$$y - y_1 = 2x_1(x - x_1)$$

or

$$y = 2x_1 x - x_1^2. \qquad (4.198)$$

The curve $y = x^2$ is convex ($y'' = 2 > 0$), that is it lies above its tangent line everywhere. Thus we can write the inequality

$$x^2 \ge 2x_1 x - x_1^2 \qquad (4.199)$$

for all x_1, with equality when $x = x_1$. Hence we write†

$$x^2 = \max_{x_1}[2x_1x - x_1{}^2].$$ (4.200)

Now we shall utilize this result on the Riccati equation

$$v' = -v^2 - p(t)v - q(t)$$ (4.201)

by replacing v^2 by its equivalent expression

$$\max_u[2uv - u^2].$$ (4.202)

Thus Eq. (4.201) now reads ‡

$$v' = -\max_u[2uv - u^2] - p(t)v - q(t)$$

$$= \min_u[u^2 - 2uv] - p(t)v - q(t)$$

$$= \min_u[u^2 - 2uv - p(t)v - q(t)].$$ (4.203)

Were it not for the minimization (or maximization) operation, Eq. (4.203) would be linear. As it is Eq. (4.203) possesses certain properties associated with linear equations. It is for these reasons that Bellman and Kalaba [72] coined the phrase *quasilinearization*.

With $u(t)$ a fixed function of t, the comparison linear equation to Eq. (4.203) is

$$w' = u^2 - 2uw - p(t)w - q(t)$$ (4.204)

where $w(0) = v(0) = c$. The solution of Eq. (4.204) is well known to be

$$w = \exp\left\{-\int_0^t [2\dot{u}(s) + p(s)]\, ds\right\}$$

$$\times \left\{c + \int_0^t [u^2(s) - g(s)] \exp\left[\int_0^s (2u(r) + p(r))\, dr\right] ds\right\}$$ (4.205)

which we denote by $w = T(f, g)$ with $f = -2u - p$, $g = u^2 - q$. Note that the positivity of the exponential allows us to conclude that $g_1(t) \geq g_2(t)$

---

† This result is immediately establishable in many other ways.

‡ The last step follows from the observation that the last two terms are independent of u, over which the minimization occurs.

for $t \geq 0$ implies $T(f, g_1) \geq T(f, g_2)$ for $t \geq 0$. We shall find this a most important property in the following work.

Suppose the solution of Eq. (4.197) exists for all t in $0 \leq t \leq t_0$. This is not a trivial assumption since the solution of a Riccati equation may have singularities. An example is $u' = 1 + u^2$, $u(0) = 0$ whose solution is $u = \tan t$. $u[(2n + 1)\pi/2]$, n an integer, does not exist!

We now wish to establish the inequality

$$w \geq v. \tag{4.206}$$

Equation (4.203) is equivalent to the inequality

$$v' \leq u^2 - 2uv - p(t)v - q(t)$$

for all $u(t)$ or

$$v' = u^2 - 2uv - p(t)v - q(t) - r(t) \tag{4.207}$$

where $r(t) \geq 0$ for $t \geq 0$. Obviously $r(t)$ depends upon u and v but this is of no consequence at the moment. In terms of the operation notation $T(f, g)$ we can write the solution to the linear Eq. (4.207) as

$$\begin{aligned} v &= T[-2u - p(t), u^2 - q(t) - r(t)] \\ &\leq T[-2u - p, u^2 - q] = w \end{aligned} \tag{4.208}$$

where the last inequality follows from the positivity property of T. Since this inequality holds for all $u(t)$, with equality for $u = v$ we have the result: *For $0 \leq t \leq t_0$, where t_0 depends upon the interval of existence of $v(t)$, we have the representation*

$$v(t) = \min_u T\left(-2u - p, u^2 - q\right) \tag{4.209}$$

where T is given by Eq. (4.205).

Thus by applying quasilinearization we have constructed an analytic expression for the Riccati equation.

In a sequence of papers Calogero [73–74–75] has used the representation Eq. (4.209) to find upper and lower bounds for $v(t)$ by appropriate choices of $u(t)$. See also Bellman and Kalaba [72] for additional references.

We are now in a position to apply the preceding theorem to generate a sequence of approximations to the solution $v(t)$ of the Riccati equation. The minimum in Eq. (4.209) is attained by the solution $v(t)$ itself. Thus if v_0 is a reasonable initial approximation we suspect that the recurrence relation

$$\left.\begin{aligned}v'_{n+1} &= v_n^2 - 2v_n v_{n+1} - p(t)v_{n+1} - q(t) \\ v_{n+1}(0) &= c\end{aligned}\right\} \qquad (4.210)$$

generates a convergent sequence by analogy with the Newton–Raphson sequence. Generally, for the differential equation $v' = g(v, t)$ the approximation scheme would be

$$\left.\begin{aligned}v'_{n+1} &= g(v_n, t) + (v_{n+1} - v_n)g_v(v_n, t), \\ v_{n+1}(0) &= c.\end{aligned}\right\} . \qquad (4.211)$$

In these two cases the recurrence relations are exactly the same scheme as would be obtained if we applied the Newton–Raphson–Kantorovich technique. However, the two approaches *are not* equivalent despite their coincidence in special cases (see Bellman and Kalaba [72]).

The sequence $\{v_n\}$ defined by Eq. (4.210) is shown to have monotone quadratic convergence (see [72]). This results from the properties of the operator $T(f, g)$ and the fact that the minimum of the right hand side of Eq. (4.210) is achieved for $v_{n+1} = v_n$.

4.15 Applications of Quasilinearization

The fundamental source, Bellman and Kalaba [72], contains a wide variety of applications and references both within our area of specialization and elsewhere. Some applications are given below.

A. $u'' = f(u), \; u(0) = u(b) = 0$ \qquad (4.212)

Let $u_0(x)$ be some initial approximation and consider the sequence $\{u_n\}$ defined by

$$\left.\begin{aligned}u''_{n+1} &= f(u_n) + (u_{n+1} - u_n)f'(u_n) \\ u_{n+1}(0) &= u_{n+1}(b) = 0.\end{aligned}\right\} \qquad (4.213)$$

This sequence is well defined for x on a sufficiently small interval and it converges monotonically and quadratically.

A typical example occurs in a heat conduction-chemical reaction problem (see Chapter 2)

$$u'' = e^u, \qquad u(0) = u(b) = 0. \qquad (4.214)$$

For $b = 1$ the analytic solution to Eq. (4.214) is

$$u(x) = -\ln 2 + 2 \ln[C \sec\{C(x - \tfrac{1}{2})/2\}] \qquad (4.215)$$

where C is the root of $\sqrt{2} = C \sec(C/4)$ which lies between 0 and $\pi/2$ ($C \approx 1.33\,60\,56$). In this case the sequence of approximations is

$$u_0 = 0$$
$$u_{n+1}'' = \exp(u_n) + (u_{n+1} - u_n)\exp(u_n)$$
$$u_{n+1}(0) = u_{n+1}(1) = 0.$$

Bellman and Kalaba [72] find substantial agreement (5 decimal places) of u_2 to the exact solution Eq. (4.215) where the computation is accomplished by a Runge–Kutta procedure (see Chapter 5).

B. GENERAL SECOND ORDER DIFFERENTIAL EQUATION

Let us consider the nonlinear system with nonlinear boundary conditions

$$\left.\begin{aligned} u'' &= f(x, u, u') \\ g_1[u(0), u'(0)] &= 0 \\ g_2[u(b), u'(b)] &= 0. \end{aligned}\right\} \qquad (4.216)$$

Quasilinearization must be applied to both the equation and the boundary conditions. Our sequence is then

$$u_{n+1}'' = \frac{\partial f}{\partial u'}(x, u_n, u_n')(u_{n+1}' - u_n') + \frac{\partial f}{\partial u}(x, u_n, u_n')(u_{n+1} - u_n) \qquad (4.217)$$

with the linearized boundary conditions

$$\frac{\partial g_1}{\partial u'}[u_n(0), u_n'(0)][u_{n+1}'(0) - u_n'(0)]$$

$$+ \frac{\partial g_1}{\partial u}[u_n(0), u_n'(0)][u_{n+1}(0) - u_n(0)] = 0 \qquad (4.218)$$

and a similar condition derived for g_2 at $x = b$.

Extensions to more general boundary conditions are also possible.

C. Higher Order Equations

Let us examine the quasilinearization process as applied to the equation

$$u^{(4)} + ke^u = 0$$

with the boundary conditions

$$u(0) = u'(0) = 0$$
$$u''(1) = u'''(1) = 0.$$

The recurrence relation becomes

$$u_{n+1}^{(4)} = -k \exp(u_n) - k(u_{n+1} - u_n) \exp(u_n)$$
$$u_{n+1}(0) = u'_{n+1}(0) = u''_{n+1}(1) = u'''_{n+1}(1) = 0. \tag{4.219}$$

By means of an Adams–Moulton integration procedure applied to Eq. (4.219) convergence to 6 decimal places occurred after four iterations.

For a large number of further applications in the calculus of variations, partial differential equations, dynamic programming etc the reader is urged to consult Bellman and Kalaba [72].

4.16 Other Methods of Approximation

No modest volume could include all of the myriad methods for obtaining approximate solutions to nonlinear ordinary differential equations. We have discussed in detail those, which in our opinion, are the most useful for problems in transport phenomena. For the sake of completeness we briefly discuss some additional methods, where they have found application and where they may be found in the literature.

A. Method of Reversion†

This method, except for the details, is essentially the same as the regular perturbation procedure. The difference is that this method *provides a set of formulae by means of which the various steps of the process can be carried out.* Pipes [13] has a variety of examples including two examples of radiative heat transfer.

---

† See Cunningham [27] and Pipes [13].

B. Method of Van der Pol[†]

This procedure is similar to the method of variation of parameters (Cunningham [27]) and to the technique of Kryloff and Bogoliuboff [76]. Lefschetz (see Andronow and Chaikin [77]) has justified some questionable steps in these procedures. Let us describe the Van der Pol method for

$$\frac{d^2x}{dt^2} + x = \mu f\left(x, \frac{dx}{dt}\right)$$

rewritten in the form

$$\frac{dx}{dt} = y, \qquad \frac{dy}{dt} = -x + \mu f(x, y). \tag{4.220}$$

When $\mu = 0$ the solution of Eq. (4.220) is

$$x = a \cos t + b \sin t, \qquad y = -a \sin t + b \cos t \tag{4.221}$$

where a and b are constants. Van der Pol obtains an approximate solution by a method similar to the method of variation of parameters in the linear theory. That is to say we assume $a = a(t)$, $b = b(t)$ so that

$$x = a(t) \cos t + b(t) \sin t, \qquad y = -a(t) \sin t + b(t) \cos t. \tag{4.222}$$

Upon substituting Eq. (4.222) into Eq. (4.220) and solving the resulting simultaneous system we obtain

$$\frac{da}{dt} = -\mu \sin t\, f[a \cos t + b \sin t, - a \sin t + b \cos t]$$

$$\frac{db}{dt} = \mu \cos t\, f[a \cos t + b \sin t, - a \sin t + b \cos t]. \tag{4.223}$$

These simultaneous equations are usually sufficiently complicated so that exact solutions are difficult to find. However if μ is zero both a' and b' are zero. Thus neither a nor b varies with time. If μ is small and f bounded it is clear that da/dt and db/dt are both $0(\mu)$ and therefore the variation of a and b in time 2π will be small. The process continues by a Fourier expansion of the right-hand side (see also Saaty and Bram [78].)

---

† See Davies and James [2].

Goodman and Sargent [79] have applied this method to the equation

$$x'' + f(x') + g(x) = h(t) \tag{4.224}$$

where $h(t)$ is a random gaussian force.

C. METHOD OF LIE SERIES†

A Lie series has the form

$$\sum_{n=0}^{\infty} \frac{t^n}{n!} D^n f(Z) = f(Z) + tDf(Z) + \cdots \tag{4.225}$$

where

$$D = g_1(Z)\frac{\partial}{\partial Z_1} + \cdots + g_k \frac{\partial}{\partial Z_k} \tag{4.226}$$

and the $g_k(Z)$ are analytic functions of $Z = (Z_1, \ldots, Z_k)$ in a neighborhood of the same point. t is a new variable independent of Z. These series can be shown to converge absolutely as a power series in t for all Z for which f and g_j are analytic. That is to say a Lie series is an analytic function in Z and t.

For example the operator corresponding to the equation $y'' + ay - by^3 = 0$, $a \geq 0$, is

$$D = y'\frac{\partial}{\partial y} - (ay - by^3)\frac{\partial}{\partial y'}.$$

D. TAYLOR-CAUCHY TRANSFORM‡

The Taylor–Cauchy transform of $f(\lambda)$ is

$$T_C(f) = \frac{1}{2\pi i}\int_C \frac{f(\lambda)}{\lambda^{m+1}}\, d\lambda, \qquad m = 0, 1, 2, \ldots.$$

Its applicability is to equations of the form

$$\sum_{j=0}^{n} a_j x^{(j)} + f(x, x', \ldots, x^{(n-1)}) = g(t).$$

---

† See Gröbner and Cap [80].
‡ See Ku *et al.* [81].

E. RATIONAL APPROXIMATION

The aforementioned techniques usually require extensive algebraic manipulation. In addition they and naively applied numerical integration can lead to disaster without knowledge as to the location of poles (if any) of the solution. The existence of poles of solutions to nonlinear differential equations occurs frequently enough so that a method for simultaneously obtaining both the solution and the poles location is highly desirable. Rational approximations can be used to reflect accurately the positions of both poles and zeros and the global behavior of the solution.

Work in this area for nonlinear equations is underway. A major tool is the theory of continued fractions (see Wall [82] or the highly readable Davenport [83]). The Riccati equation has been treated by this method in the work of Merkes and Scott [84] and Fair [85]. Fair and Luke [86] develop the method for the generalized second order Riccati equation,

$$(A_0 + B_0 y)y'' + (C_0 + D_0 y)y' - 2B_0(y')^2 + E_0 + F_0 y + G_0 y^2 + H_0 y^3 = 0$$
$$(4.227)$$

with $y(0) = \alpha_0$, $y'(0) = \beta_0$, $\alpha_0 \beta_0 \neq 0$.

This method is very promising and warrants extensive effort.

REFERENCES

1. Liénard, A., *Rev. Gen. Elec.* **37**, 901 (1928).
2. Davies, T. V., and James, E. M., "Nonlinear Differential Equations," Addison-Wesley, Reading, Massachusetts, 1966.
3. Goursat, E. V., Hedrik, E. R., and Dunkel, O., "A Course in Mathematical Analysis," Vol. II, Part II, *"Differential Equations."* Ginn, Boston, Massachusetts, 1917.
4. Ince, E. L., "Ordinary Differential Equations," Dover, New York, 1956.
5. McLachlan, N. W., "Ordinary Nonlinear Differential Equations," 2nd ed. Oxford Univ. Press, London and New York, 1956.
6. Thomas, G. B., Jr., "Calculus and Analytic Geometry," Addison-Wesley, Reading, Massachusetts, 1953.
7. Hildebrand, F. B., "Introduction to Numerical Analysis," McGraw-Hill, New York, 1956.
8. Hamming, R. W., "Numerical Methods for Scientists and Engineers," McGraw-Hill, New York, 1962.
9. Todd, J. ed., "Survey of Numerical Analysis," McGraw-Hill, New York, 1962.

10. "Tables of Chebyshev Polynomials," Appl. Math. Ser. **9**, Natl. Bur. St. U. S. Govt. Printing Office, Washington, D.C., 1952.

11. Lanczos, C., *J. Math. and Phys.* **17**, 123 (1938).

12. Pipes, L. A., "Applied Mathematics for Engineers and Physicists," 2nd ed. McGraw-Hill, New York, 1958.

13. Pipes, L. A., Operational methods in nonlinear mechanics, Rep. #51-10 December 1951, Dept. of Eng., Univ. of California, Los Angeles, California, 1951.

13a. Picard, E., "Traité d'Analyse," Paris, 1896.

14. Nagaoka, H., *Proc. Imp. Acad. (Tokyo)* **3**, 13 (1927).

15. Poincaré, H., "Mecanique Celeste," Vol. I, Gauthier-Villars, Paris, 1892.

16. Shaefer, C., *Ann. Physik* **33**, 1216 (1910).

17. Volterra, V., "Lecons sur les Équationes Intégrales," Gauthier-Villars, Paris, 1913.

18. Lock, G. S. H., *Bull. Mech. Eng. Educ.* **5**, 71 (1966).

19. Weyl, H., *Proc. Natl. Acad. Sci U.S.* **27**, 578 (1941); **28**, 100 (1942).

20. Abramowitz, M., *J. Math. and Phys.* **30**, 162 (1951).

21. Davies, T. V., *Proc. Roy. Soc.* **A273**, 496, 518 (1963).

22. James, E. M., *Proc. Roy. Soc.* **A273**, 509 (1963).

23. Nowinski, J. L., and Ismail, I. A., *in* "Developments in Theoretical and Applied Mechanics" (N. A. Shaw, ed.), Vol. II, p. 35. Pergamon Press, Oxford, 1965.

24. Ames, W. F., and Sontowski, J. L., *J. Appl. Mech.* **33**, 218 (1966).

25. Stoker, J. J., "Nonlinear Vibrations in Mechanical and Electrical Systems," Wiley (Interscience), New York, 1950.

26. McLachlan, N. W., "Ordinary Nonlinear Differential Equations in Engineering and Physical Sciences," 2nd ed. Oxford Univ. Press, London and New York. 1955.

27. Cunningham, W. J., "Introduction to Nonlinear Analysis," McGraw-Hill, New York, 1958.

28. Kryloff, N. M., and Bogoliuboff, N. N., "Introduction to Nonlinear Mechanics," Princeton Univ. Press, Princeton, New Jersey, 1943.

29. Bellman, R. E., "Perturbation Techniques in Mathematics, Physics and Engineering," Holt, New York, 1964.

29a. Linsted, E. (See Struble, R. A., "Nonlinear Differential Equations," McGraw-Hill, New York, 1962).

30. Bautin, N. N., *Mat. Sb.* (N.S.) **30**, 72, 181 (1952).

31. Schlichting, H., "Boundary Layer Theory," 4th ed. (transl. by J. Kestin). McGraw-Hill, New York, 1960.

32. Shohat, J., *J. Appl. Phys.* **15**, 568 (1944).

33. Bellman, R., *Quart. Appl. Math.* **13**, 195 (1955).

34. Bellman, R., Paper 55-APM-33. Am. Soc. Mech. Eng., New York, 1955.

34a. Van der Pol, B., *Phil. Mag.* **2** (7), 978 (1926).

35. Urabe, M., *IRE Trans.* PGCT-7, 382 (1960).

36. Friedrichs, K. O., Theory of viscous fluids, *In* "Fluid Dynamics," Chapter 4. Brown Univ. Press, Providence, Rhode Island, 1942.

37. Bretherton, F. P., *J. Fluid Mech.* **12**, 591 (1962).
38. Friedrichs, K. O., "Special Topics in Fluid Mechanics," p. 126. N.Y.U. Press, New York, 1953.
39. Friedrichs, K. O., "Special Topics in Analysis," p. 184. N.Y.U. Press, New York, 1954.
40. Kaplun, S., *Z. Angew. Math. Phys.* **5**, 111 (1954).
41. Kaplun, S., and Lagerstrom, P. A., *J. Math. Mech.* **6**, 585 (1957).
42. Lagerstrom, P. A., *J. Math. Mech.* **6**, 605 (1957).
43. Lagerstrom, P. A., and Cole, J. D., *J. Ratl. Mech. Anal.* **4**, 817 (1955).
44. Van Dyke, M., "Perturbation Methods in Fluid Mechanics," Academic Press, New York, 1964.
45. Wasow, W., "Asymptotic Expansions for Ordinary Differential Equations," Wiley (Interscience), New York, 1965.
46. Wasow, W., *Commun. Pure Appl. Math.* **9**, 93 (1956)..
47. Erdelyi, A., *Bull. Am. Math. Soc.* **68**, 420 (1962).
48. Harris, W. A., Jr., *Duke Math. J.* **29**, 429 (1962).
49. Levin, J. J., and Levinson, N., *J Ratl. Mech. Anal.* **3**, 247 (1954).
50. Lighthill, M. J., *Phil. Mag.* [7], **40**, 1179 (1949).
51. Lin, C. C., *J. Math. and Phys.* **33**, 117 (1954).
52. Lighthill, M. J., *Z. Flugwiss.* **9**, 267 (1961).
53. Wasow, W., *J. Ratl. Mech. Anal.* **4**, 751 (1955).
54. Sandri, G., *Ann. Phys. (N.Y.)* **24**, 332, 380 (1962).
55. Sandri, G., *Nuovo Cimento* **36**, 67 (1965).
56. Sandri, G., *In* "Nonlinear Partial Differential Equations—Methods of Solution" (W. F. Ames, ed.). Academic Press, New York, 1967.
57. Bowen, J. R., Acrivos, A., and Oppenheim, A. K., *Chem. Eng. Sci.* **18**, 177 (1963).
58. Free, E. A., Ph.D. Thesis, Univ. of Wales, Aberystwyth, Wales, 1963.
59. Carrier, G. F., and Lewis, J. A., *Advan. Appl. Mech.* **3**, 12 (1953).
60. Galerkin, B. G., *Vestn. Inzh. i Tekhn.* p. 879 (1915).
61. Kantorovich, L. V., and Krylov, V. I., "Approximate Methods of Higher Analysis," Wiley (Interscience), New York, 1958.
62. Collatz, L., "The Numerical Treatment of Differential Equations," (English ed.) Springer, Berlin, 1960.
63. Crandall, S. H., "Engineering Analysis," McGraw-Hill, New York, 1956.
64. Ames, W. F., "Nonlinear Partial Differential Equations in Engineering," Academic Press, New York, 1965.
65. Hetenyi, M., "Beams on Elastic Foundations," p. 60, Univ. of Michigan Press, Ann Arbor, Michigan, 1946.
66. Frazer, R. A., Jones, W. P., and Skan, S. W., Approximations to functions and to the solutions of differential equations, Rept. and Mem. No. 1799, Aeron. Res. Comm., 1937.
67. Bickley, W. G., *Phil. Mag.* [7], **32**, 50 (1941).
68. Finlayson, B. A., and Scriven, L. E., *Appl. Mech. Rev.* **19**, 735, (1966).

69. Finlayson, B. A., and Scriven, L. E., *Intern. J. Heat Mass Trans.* **10**, 799 (1967).
70. Jain, M. K., *Appl. Sci. Res. Sect. A* **11**, 177 (1963).
71. Collatz, L., *Z. Angew. Math. Mech.* **33**, 118 (1953).
72. Bellman, R. E., and Kalaba, R. E., "Quasilinearization and Nonlinear Boundary-Value Problems," American Elsevier, New York, 1965.
73. Calogero, F., *Nuovo Cimento* **27**, 261 (1963).
74. Calogero, F., *J. Math. Phys.* **4**, 427 (1963).
75. Calogero, F., *Nuovo Cimento* **28**, 320 (1963).
76. Kryloff, N., and Bogoliuboff, N., "Introduction to nonlinear mechanics" (Ann. Math. Studies, Vol. II) Princeton Univ. Press, Princeton, New Jersey, 1943.
77. Andronow, A., and Chaikin, C. E., "Theory of Oscillations," Princeton Univ. Press, Princeton, New Jersey, 1949.
78. Saaty, T. L., and Bram, J., "Nonlinear Mathematics," McGraw-Hill, New York, 1964.
79. Goodman, T. R., and Sargent, T. P., Launching of airborne missiles underwater, "Effect of Nonlinear Submarine Roll Damping on Missile Response in Confused Seas," Part XI. Allied Research Associates, Inc., Boston, Massachusetts, 1961.
80. Gröbner, W., and Cap, F., A new method to solve differential equations: Lie series and their applications in physics and engineering, Rept. N62558-2992, Office Naval Res. Washington, D. C., August 1962.
81. Ku, Y. H., Wolf, A. A., and Dietz, J. H., *Trans. AIEE* **79** (II), 183 (1960) (see also Ku, Y. H., *J. Franklin Inst.* **271**, 108 (1961).
82. Wall, H. S., "Continued Fractions," Van Nostrand, Princeton, New Jersey, 1948.
83. Davenport, H., "The Higher Arithmetic," Hutchinson's Univ. Library, London, 1952.
84. Merkes, E. P., and Scott, W. T., *J. Math. Anal. Appl.* **4**, 309 (1962).
85. Fair, W. G., *Math. Computat.* **18**, 627 (1964).
86. Fair, W. G., and Luke, Y. L., Rational approximation to the generalized Duffing equation, *J. Nonlinear Mech.* **1**, 209 (1966).

5

NUMERICAL METHODS

Introduction

No single numerical method is applicable to every ordinary differential equation or for that matter to every member of the much smaller class of ordinary linear differential equations. The field is immense—research continues and at an accelerating rate. Since there are a large variety of problem types we need a variety of methods, each most appropriate to its particular and relatively small class of problems. This point cannot be overemphasized because many computer "libraries" have only one or two routines for ordinary differential equations (the most popular are the Adams and Runge–Kutta methods.) There is therefore a tendency among the less experienced to apply the "canned" program on *all* problems. This is done even though better methods may be available.

The large variety of possible problems precludes a well defined useful choice criterion. The linear equations have simplifications (not applicable to the nonlinear)—notably the possibility of superposition of partial solutions such as the familiar linear combination of complementary solutions and particular integrals. Other factors, chiefly the auxiliary conditions, may have great importance on the selection of a method. If these are specified at a single value of the independent variable then we have an *initial value or propagation problem*. The numerical methods for such problems need to be fundamentally different† from those for *boundary value or steady state problems*. In the latter case the auxiliary conditions are

---

† Of course this same statement is true for analytic and approximate procedures as well.

shared among two or more points in the range of the independent variable. In many cases nonlinear initial value problems require less coding, storage, and computation than linear boundary value problems.

Yet a third consideration is the form of the equation under consideration. If various methods, such as quasilinearization, apply then that information should be factored into our selection of a technique. If the nonlinear initial value problem has an equation like

$$y'' = f(x, y) \tag{5.1}$$

it is probably better to use a finite difference method, while for the equation

$$y'y'' + y \sin y' = f(x, y) \tag{5.2}$$

one is advised to develop two similtaneous first order equations and use a Runge–Kutta method.

Lastly the form of the solution and its asymptotic behavior in an infinite range of the independent variable may dictate the choice of method. In such cases, for initial value problems, the question of *stability* arises. We do not wish our method to generate *spurious solutions* which may grow without bound and swamp the *true solution*. The type of difficulty one is likely to face is amply illustrated by the simple example

$$\frac{dx}{dt} = y, \qquad\qquad x(0) = 1$$

$$\frac{dy}{dt} = 10b^2 x + 9by, \qquad y(0) = -b \tag{5.3}$$

in which $b > 0$. Its general solution is

$$x = C_1 e^{-bt} + C_2 e^{10bt}$$
$$y = -bC_1 e^{-bt} + 10bC_2 e^{10bt} \tag{5.4}$$

and the particular solution satisfying the initial conditions is obtained for $C_1 = 1$, $C_2 = 0$. No matter what numerical method is used we are incapable of working with an infinite decimal. Thus, due to round-off errors, a C_2 component will be introduced. If at $t = \eta$ this component is λ, it will be λe^{20b} at $\eta + 2$. Ultimately this "error" will totally overshadow the desired solution and lead to entirely erroneous results, regardless of the number of decimals carried in the computation.

In a particular problem the picture may not be this dark. Yet this example is a warning. Careful analysis should always precede the initiation of computation. Numerical integration should not be undertaken when no definite information about the solution is available.†

Our considerations in this chapter will be the numerical solution of initial and boundary value problems. These computations are accomplished on two basic species of computing machines—the *analog* kind and the *digital* kind. Both perform their functions in ways which involve symbols for or representations of numbers, but each of the two kinds uses a fundamentally different sort of symbol—which is why they are classified as different kinds of machines.

In an analog computer, a number is represented by the result of measuring some quantity; e.g., the voltage on a line, the length of a stick, the amount of rotation of a wheel. The familiar slide rule employs the analog method of number representation, using various lengths marked off along a piece of wood, or around the edge of a circular disc, for that purpose. The speedometer on your car represents the number denoting the rate at which the road is passing under the wheels analogically, but represents the number of miles the car has travelled in a digital manner; i.e., by counting in terms of a number of positions, in each of which one of ten different kinds of mark is able to appear. There are, of course, other ways than this one of representing a number that results from counting, but this particalur way has proved more useful than any other, and this way of representing numbers distinguishes digital computing machines as a class.

A digital computer invented solely for the purposes of exposition has been used for a number of years by Ames and Robinson [1].

Since there are basically two kinds of computers‡ there will exist a need for two distinct approaches. In the first few sections of this chapter we describe digital methods and give some examples. The chapter is completed with a discussion of analog techniques and examples.

Among the books specifically devoted to the numerical solution of differential equations are Bennett *et al.* [2], Collatz [3], Fox [4, 5], Levy and Baggott [6], Milne [7], Mikeladze [8], von Sanden [9], and Henrici [10, 11]. Various other works pay substantial attention to this area – we

---

† One analyst described this with the phrase "The most practical thing a practical man can do is think theoretically."

‡ A third "kind" results from a hybridization of these.

mention those works of Todd [12] which contains over 100 additional references and Hildebrand [13]. Research in digital "discrete" variable methods is described in a variety of journals. Most frequently mentioned journals include: *Journal of the Association for Computing Machinery, Mathematics of Computation, Journal of the Society for Industrial and Applied Mathematics* (Series *C* on Numerical Analysis,) *Numerische Mathematik, Journal of Computational Physics*, and the *Computer Journal*.

Among the books specifically devoted to analog computer methods and technology we find the works of Johnson [14], Korn and Korn [15], Karplus and Soroka [16], Warfield [17], Scott [18], and Jackson [19].

The most versatile and accurate of the analog computers is the electronic analog. Because of the relative ease of construction analog computers are usually cheaper than digital devices. In addition analogs are frequently faster than the digital in solving complex problems because interconnected analog devices can perform many operations simultaneously.

In view of the above advantages of analogs why have digital computers achieved such tremendous importance? — speaking, of course, from a technical (nonbusiness) viewpoint. The two important reasons usually quoted are their *versatility* and the *greater precision* they can attain. The sequential nature of digital computation affords a high degree of flexibility. Analog computers of reasonable cost are limited to a precision of 10^{-4} to 10^{-5} and these devices are subject to nonanalyzable error due to noise. Digital computation has no theoretical precision limit. Further, truncation and round-off errors are subject to mathematical analysis. Thus analog computers are widely used when high precision is unnecessary or unwarranted.

5.1 Finite Differences

All finite difference formulae are based upon polynomial approximation. They give exact results when operating upon a polynomial of the proper degree. In all other cases the formulae are approximations usually expressed as series of finite differences. We must of necessity use only a finite number of terms of these series so we wish the *truncation error* to be small.

The following notations for various differences have now become standard. We shall think of the finite differences as being applied to a function

$y = f(x)$, with a constant interval size $h = x_{n+1} - x_n$ and we denote by y_n that value of f at x_n—that is $y_n = f(x_n)$. Now define the following operators:

Forward difference:	$\Delta y_n = y_{n+1} - y_n$	(5.5a)
Backward difference:	$\nabla y_n = y_n - y_{n-1}$	(5.5b)
Central difference:	$\delta y_n = y_{n+1/2} - y_{n-1/2}$	(5.5c)
Averaging operator:	$\mu y_n = \frac{1}{2}[y_{n+1/2} + y_{n-1/2}]$	(5.5d)
Shift operator:	$E y_n = y_{n+1}$	(5.5e)
Integral operator:	$Jy = \displaystyle\int_x^{x+h} y(t)\, dt$	(5.5f)
Differential operator:	$Dy = dy/dx$	(5.5g)

For most purposes the operators of Eq. (5.5) can be manipulated according to the ordinary laws of algebra. The general theory and exceptions to these laws are discussed by Hildebrand [13], Jordan [20] or Milne-Thompson [21]. The formal manipulation of these operators is used to produce finite difference expressions. Thus we have

$$DJ = JD = \Delta = E - 1,$$
$$E = 1 + \Delta, \qquad (5.6)$$
$$\nabla = 1 - E^{-1}, \text{ etc.}$$

From Eq. (5.6) we have for integral p

$$E^p y_n = y(x_n + ph) = (1 + \Delta)^p y_n = \left[1 + p\Delta + \binom{p}{2}\Delta^2 + \cdots\right] y_n$$

$$= \left[y_n + p(\Delta y_n) + \binom{p}{2}(\Delta^2 y_n) + \cdots\right] \qquad (5.7)$$

where we have used $\Delta^2 = \Delta\Delta$ and $\binom{p}{j} = \dfrac{p!}{j!(p-j)!}$.

Equation (5.7) is a finite difference formula for calculating $y(x_n + ph)$ in terms of various finite differences at y_n.

The averaging operator is required because the values $y_{n+1/2}$ and $y_{n-1/2}$

do not appear in tables of values. Similarily odd central differences $\delta^{2p+1}y_n$ are lacking. Some additional operational identities are useful:

$$\mu = \tfrac{1}{2}(E^{1/2} + E^{-1/2}), \qquad \delta = E^{1/2} - E^{-1/2},$$

$$\mu^2 = 1 + \tfrac{1}{4}\delta^2, \qquad\qquad E = \frac{1}{1 - \nabla}. \qquad (5.8)$$

Thus for example we have

$$1 = \mu\mu^{-1} = \mu(1 + \tfrac{1}{4}\delta^2)^{-1/2}$$

$$= \mu[1 - \tfrac{1}{8}\delta^2 + \frac{3}{128}\delta^4 - \cdots]$$

providing a method of introducing the averaging operator.

Our primary purpose is to obtain formulae useful in approximating derivatives. To do so we need to relate the operator D to the other operators. For this purpose we notice that the Taylor series expansion

$$f(x + h) = f(x) + \frac{h}{1!}f'(x) + \frac{h^2}{2!}f''(x) + \cdots + \frac{h^n}{n!}f^{(n)}(x) + \cdots \qquad (5.9)$$

can be written in the operational form

$$Ef(x) = \left[1 + \frac{hD}{1!} + \frac{h^2D^2}{2!} + \cdots + \frac{h^nD^n}{n!} + \cdots\right]f(x). \qquad (5.10)$$

The series in the brackets of Eq. (5.10) is the expansion of e^{hD}. Thus we deduce the curious relation between the operator E and D that

$$E = e^{hD}. \qquad (5.11)$$

Since our approximations are based upon polynomials we interpret this statement to mean that the operators E and $1 + hD/1! + h^2D^2/2! + \cdots + (hD)^n/n!$ are equivalent when applied to any polynomial $p_n(x)$ of degree n, for any n.†

By formal operations we can develop the additional relations

$$hD = \log E = \log(1 + \Delta) = -\log(1 - \nabla)$$

$$= 2 \sinh^{-1}\frac{\delta}{2}$$

$$= 2 \log[(1 + \tfrac{1}{4}\delta^2)^{1/2} + \tfrac{1}{2}\delta]. \qquad (5.12)$$

---

† In general this is what we mean by any statement equating operators.

Reverting to our previous notation we have as a result of Eq. (5.12) for derivatives at pivotal points

$$y_r' = \frac{1}{h}\left[\Delta - \frac{1}{2}\Delta^2 + \frac{1}{3}\Delta^3 - \cdots \right] y_r. \tag{5.13}$$

This expression uses the *forward differences* at the relevant point. In terms of the forward differences at the previous point we can also write

$$y_r' = \frac{1}{h}\left[\Delta + \frac{1}{2}\Delta^2 - \frac{1}{6}\Delta^3 + \cdots \right] y_{r-1}. \tag{5.14}$$

By repeated operation of Eq. (5.12) there follows

$$y_r^{(k)} = \frac{1}{h^k}\left[\log(1 + \Delta)\right]^k y_r$$

$$= \frac{1}{h^k}\left[\Delta^k - \frac{k}{2}\Delta^{k+1} + \frac{k(3k+5)}{24}\Delta^{k+2} - \frac{k(k+2)(k+3)}{48}\Delta^{k+3} + \cdots \right] y_r \tag{5.15}$$

which is a forward difference formula for the kth derivative.

Central differences are valuable in a number of situations. Thus we find

$$y_r' = \frac{2}{h}\sinh^{-1}\delta/2\, y_r = \frac{2\mu \sinh^{-1}\delta/2}{h(1 + \delta^2/4)^{1/2}}\, y_r$$

$$= \frac{\mu}{h}\left(\delta - \frac{1}{3!}\delta^3 + \frac{1}{30}\delta^5 - \cdots \right) y, \tag{5.16}$$

that is, a *mean odd central difference*. With central differences we also have a simple useful result for the derivative at a half-way point

$$y_{r+1/2}' = \left(\delta - \frac{1}{24}\delta^3 + \frac{3}{640}\delta^5 - \cdots \right) y_{r+1/2} \tag{5.17}$$

and its generalization

$$y_{r+1/2}'' = \mu\left(\delta^2 - \frac{5}{24}\delta^4 - \frac{259}{5760}\delta^6 - \cdots \right) y_{r+1/2}. \tag{5.18}$$

For second derivatives we also have

$$y_r'' = \frac{1}{h^2}\left(\Delta^2 - \frac{1}{12}\Delta^4 + \frac{1}{12}\Delta^5 - \cdots \right) y_{r-1} \tag{5.19}$$

and the central difference expression

$$y_r'' = \frac{1}{h^2}\left(\delta^2 - \frac{1}{12}\delta^4 + \frac{1}{90}\delta^6 - \cdots\right)y_r.$$ (5.20)

Further formulae and a detailed discussion of their derivation is available in Hildebrand [13, Chapter 5].

All methods for developing the numerical solution of differential equations yield for an abscissa sequence $\{x_n\}$, a sequence $\{y_n\}$ of solutions which approximate the true solution. Calculational methods are either *one step* or *multistep methods*. The most famous representative of the one step group is the *Runge–Kutta* generalization of Simpson's rule [22, 23]. The classical example of the multistep method is that of Bashforth and Adams [24].

In what follows we shall discuss the development of numerical techniques for the simple equation

$$\frac{dy}{dx} = f(x, y), \qquad y(x_0) = y_0$$ (5.21)

and then give the generalizations.

5.2 One Step Methods

Every one step method is expressible in the form

$$y_{n+1} = y_n + h_n\phi(x_n, y_n; h_n), \qquad n = 0, 1, 2, \ldots$$ (5.22)

where $\phi(x, y; h)$ is some function and $h_n = x_{n+1} - x_n$ (we do not assume $h_n = h_{n+1}$ for all n). The choice of ϕ should be "reasonable" in the sense that for fixed (x, y) in the integration domain R we wish

$$\phi(x, y; h) \to f(x, y) \qquad \text{as} \qquad h \to 0.$$ (5.23)

Let $y(x)$ be the exact solution of Eq. (5.21) and set

$$T(x, h) = y(x) + h\phi(x, y(x); h) - y(x + h)$$ (5.24)

which is the *truncation* error at the point x. For fixed x in $x_0 \le x \le \bar{x}$ we have

$$T(x, h) = o(h) \qquad \text{as} \qquad h \to 0.$$

If p is the largest integer such that

$$T(x, h) = O(h^{p+1}) \qquad \text{as} \qquad h \to 0 \qquad (5.25)$$

then p is called the *order* of the method.

The simplest choice of ϕ satisfying Eq. (5.23) is $\phi(x, y; h) = f(x, y)$ and the corresponding numerical method is

$$y_{n+1} = y_n = h_n f(x_n, y_n) \qquad (5.26)$$

a procedure proposed by Euler [25]. This method of Euler is of *first order* if $f(x, y)$ has continuous first partial derivatives since

$$
\begin{aligned}
T(x, h) &= y(x) + hf[x, y(x)] - y(x + h) \\
&= y(x) + hy'(x) - y(x + h) = O(h^2).
\end{aligned}
\qquad (5.27)
$$

The last result is easily developed by expanding $y(x + h)$ in a Taylor series about x. This result also demonstrates that the Euler method utilizes the first two terms of Taylor series.

A natural extension of Euler's method is the *method of Taylor series* which uses the first $p + 1$ terms of the series. As we shall see we are then lead to a method of order p. Let $f(x, y) \, \varepsilon C^p$ †, $p > 1$ and set

$$f_0(x, y) = f(x, y)$$

$$f_{k+1}(x, y) = \frac{\partial f_k(x, y)}{\partial x} + \frac{\partial f_k(x, y)}{\partial y} f(x, y), \qquad k = 0, 1, 2, ..., p - 2. \quad (5.28)$$

Immediately obvious is the statement that

$$f_k\{x, y(x)\} = y^{(k+1)}(x).$$

Now select

$$\phi(x, y; h) = \sum_{k=0}^{p-1} \frac{h^k}{(k + 1)!} f_k(x, y) \qquad (5.29)$$

leading us to a method of order p since

$$T(x, h) = y(x) + h \sum_{k=0}^{p-1} \frac{h^k}{(k + 1)!} y^{(k+1)}(x) - y(x + h) = O(h^{p+1}).$$

---

† We use this notation to mean that f has continuous partial derivatives of order p.

The method of Taylor series is quite efficient for linear systems but in the more general nonlinear cases it becomes very difficult to apply because of the increasing complexity of Eq. (5.28). On the other hand it has several advantages. One is an easily applied check (with equal intervals) against both blunder and truncation error. One checks by reversing the sign of h in Eq. (5.22). This changes the signs of the odd derivatives and the resulting series should reproduce the computed values at $x_n - h$, the beginning of the previous step. A second advantage is that the Taylor series coefficients decrease faster than those of analogous finite difference formula. Thus one can use a large interval with many terms or a small interval with a few terms. Because of the economization resulting from Chebyshev series some analysts prefer to use a method based upon them (see Fox [5, Chapters 9, 10]).

Variations on the method have been developed for linear equations by Wilson [26] and for the general nonlinear equation

$$y^{(n)} = f\{x, y, y', \ldots, y^{(n-1)}\} \tag{5.30}$$

by Gibbons [27].

5.3 Runge–Kutta Methods

A means of evading the complications of successive differentiations and at the same time preserving the increased accuracy afforded by Taylor series was developed by Runge [22] and improved by Kutta [23] and Heun [28]. Kutta suggested constructing ϕ, with undetermined parameters, in the form

$$\left.\begin{aligned}
\phi(x, y; h) &= \sum_{s=1}^{r} \alpha_s k_s, \qquad k_1(x, y) = f(x, y), \\
k_s(x, y; h) &= f\left(x + \mu_s h, y + h \sum_{j=1}^{s-1} \lambda_{s-1,j} k_j\right), \quad s = 2, \ldots, r.
\end{aligned}\right\} \tag{5.31}$$

For given r, the parameters α_s, μ_s, and λ_{sj} are to be determined so that the order p of Eq. (5.31) is as large as possible.

Upon expansion of the truncation error T into a Taylor series in h we have

$$T(x, h) = \sum_{k=0}^{\infty} \frac{1}{k!} \left[\frac{\partial^k T(x, h)}{\partial h^k}\right]_{h=0} h^k. \tag{5.32}$$

From Eq. (5.24) we see that $T(x, 0) = 0$ and that

$$\left[\frac{\partial^k T(x, h)}{\partial h^k}\right]_{h=0} = k\left[\frac{\partial^{k-1}\phi}{\partial h^{k-1}}\right]_{h=0} - y^{(k)}(x), \qquad (k > 0).$$

Hence Kutta's method is of order p if $f \varepsilon C^p$ and

$$\frac{\partial^{k-1}\phi}{\partial h^{k-1}} - \frac{1}{k}y^{(k)}(x) \begin{cases} \equiv 0 & \text{for} \quad 1 \le k \le p, \\ \\ \not\equiv 0 & \text{for} \quad k = p + 1. \end{cases} \qquad (5.33)$$

The detailed calculations from Eq. (5.33) show that the p identities from that system are in fact a set of nonlinear equations for the parameters α_s, μ_s, and λ_{sj}. For each r, there will be a largest value of $p = p^*(r)$ for which these equations are solvable. In fact

$$p^*(r) = r \qquad \text{if} \quad 1 \le r \le 4. \qquad (5.34)$$

There is a certain degree of freedom in developing the solution of these systems. For $r = 2$ it is one and for $r = 3$ and 4 there are two degrees of freedom. With the exception of minor typographical errors the details of the general computation are available in the work of Kopal [29]. We illustrate the development of a *third order process*.

Following Eq. (5.31) we define the sequence

$$k_1 = f(x_n, y_n) \qquad (5.35a)$$
$$k_2 = f(x_n + \mu_2 h, y_n + h\lambda_{11}k_1) \qquad (5.35b)$$
$$k_3 = f(x_n + \mu_3 h, y_n + h\lambda_{21}k_1 + h\lambda_{22}k_2) \qquad (5.35c)$$

and

$$y_{n+1} = y_n + h[\alpha_1 k_1 + \alpha_2 k_2 + \alpha_3 k_3]. \qquad (5.36)$$

We wish to determine the coefficients such that the expansion of $y_{n+1} = y(x_{n+1})$ about x_n, in powers of h, agrees with the solution of $y' = f(x, y)$ to a prescribed number of terms of the Taylor series.

Expansion of Eqs. (5.35) into Taylor series in two variables gives us the relations†

$$k_1 = f(x_n, y_n)$$

$$k_2 = f(x_n, y_n) + h[\mu_2 f_x + \lambda_{11} f f_y]$$
$$\quad + h^2[\tfrac{1}{2}\mu_2^2 f_{xx} + \mu_2 \lambda_{11} f f_{xy} + \tfrac{1}{2}\lambda_{11}^2 f^2 f_{yy}] + \cdots$$

$$k_3 = f(x_n, y_n) + h[\mu_3 f_x + \lambda_{21} f f_y + \lambda_{22} f f_y] \qquad (5.37)$$
$$\quad + h^2[\tfrac{1}{2}\mu_3^2 f_{xx} + \mu_3(\lambda_{21} + \lambda_{22}) f f_{xy}$$
$$\quad + \tfrac{1}{2}(\lambda_{21} + \lambda_{22})^2 f^2 f_{yy} + \lambda_{22}(\mu_2 f_x + \lambda_{11} f f_y) f_y] + \cdots$$

where we retain terms up to h^2 [that is, h^3 in Eq. (5.36)].

Now the Taylor series technique from Eqs. (5.23), (5.28), and (5.29) gives

$$y_{n+1} = y_n + hf(x_n, y_n)$$
$$\quad + \tfrac{1}{2}h^2[f_x + ff_y] + \tfrac{1}{6}h^3[f_{xx} + 2ff_{xy} + f^2 f_{yy} + f_x f_y + ff_y^2]. \qquad (5.38)$$

We then equate the respective coefficients of all the independent derivatives of f on the right hand side of Eq. (5.38) and the expanded right hand side of Eq. (5.36). As a result we have following relations:

$$\alpha_1 + \alpha_2 + \alpha_3 = 1, \qquad \mu_2 \alpha_2 + \mu_3 \alpha_3 = \tfrac{1}{2},$$

$$\lambda_{11} \alpha_2 + (\lambda_{21} + \lambda_{22})\alpha_3 = \tfrac{1}{2}, \qquad \tfrac{1}{2}\mu_2^2 \alpha_2 + \tfrac{1}{2}\mu_3^2 \alpha_3 = \tfrac{1}{6}, \qquad (5.39)$$

$$\mu_2 \lambda_{11} \alpha_2 + \mu_3(\lambda_{21} + \lambda_{22})\alpha_3 = \tfrac{1}{3}, \qquad \tfrac{1}{2}\lambda_{11}^2 \alpha_2 + \tfrac{1}{2}(\lambda_{21} + \lambda_{22})^2 \alpha_3 = \tfrac{1}{6},$$

$$\mu_2 \lambda_{22} \alpha_3 = \tfrac{1}{6}, \qquad \lambda_{11} \lambda_{22} \alpha_3 = \tfrac{1}{6}.$$

When we examine these equations we see that

$$\mu_2 = \lambda_{11}, \qquad \mu_3 = \lambda_{21} + \lambda_{22} \qquad (5.40)$$

and there are *four* independent relations among the remaining *six* unknowns—that is, two degrees of freedom. If any two, say μ_2 and μ_3, are assigned arbitrary values the remaining parameters are uniquely determined.

---

† The subscripts in f_x, f_y denote partial derivatives, e.g., $f_y = \partial f/\partial y$. All derivatives and functions are evaluated at (x_n, y_n).

Finally, the truncation error is

$$T_3 = \frac{h^4}{4!}[\{1 - 4(\mu_2{}^3\alpha_2 + \mu_3{}^3\alpha_3)\}D^3f + (1 - 12\mu_2{}^2\lambda_{22}\alpha_3)f_y D^2 f$$
$$+ (3 - 24\mu_2\mu_3\lambda_{22}\alpha_3)Df\,D(f_y) + f_y{}^2 Df] \tag{5.41}$$

where $D = $ operator $= ((\partial/\partial x) + f\,\partial/\partial y)$. This error expression is too cumbersome to be of much use and except in fairly trivial cases it does not seem possible to choose the arbitrary parameters so the truncation error is significantly reduced.

For the most used case, that of fourth order, we again meet two degrees of freedom. μ_2 and μ_3 can again be given arbitrary values and the remaining parameters are uniquely determined. This appears to be the limiting valuable case for the inclusion of extra terms leads to greatly increased complexity at all levels. Various authors have proposed different values of the parameters thereby leading to small advantages. Three of these are compared and the results of Martin [30] on accuracy and storage requirements discussed in Table 5.1.

TABLE 5.1

	Standard Runge–Kutta	Gill's modification[a]	Kutta's 3/8 rule[b]
α_1	$\frac{1}{6}$	$\frac{1}{6}$	$\frac{1}{8}$
α_2	$\frac{1}{3}$	$\frac{1}{3}(1 - 2^{-1/2})$	$\frac{3}{8}$
α_3	$\frac{1}{3}$	$\frac{1}{3}(1 + 2^{-1/2})$	$\frac{3}{8}$
α_4	$\frac{1}{6}$	$\frac{1}{6}$	$\frac{1}{8}$
μ_2	$\frac{1}{2}$	$\frac{1}{2}$	$\frac{1}{3}$
μ_3	$\frac{1}{2}$	$\frac{1}{2}$	$\frac{2}{3}$
μ_4	1	1	1
λ_{11}	$\frac{1}{2}$	$\frac{1}{2}$	$\frac{1}{3}$
λ_{21}	0	$\frac{1}{2}(2^{1/2} - 1)$	$-\frac{1}{3}$
λ_{31}	0	0	1
λ_{22}	$\frac{1}{2}$	$1 - 2^{-1/2}$	1
λ_{32}	0	$-2^{-1/2}$	-1
λ_{33}	1	$1 + 2^{-1/2}$	1
Storage registers:	5	3	4
Rounding error guard:	Moderate	Easy	Lengthy

[a] See Gill [31],   [b] See Kutta [23].

Simple expressions for the truncation errors in the preceding formulae are not known. In practice an *estimate* of the error can be obtained as follows: Let the truncation error associated with a pth order process, in progressing from x_n to $x_{n+1} = x_n + h$, in a single step, be denoted by $E_n h^{p+1}$. Suppose E_n varies slowly with n and is nearly independent of h, for small h. Let the true solution at x_{n+1} be Y_{n+1} and let the value obtained by *two* steps starting at x_{n-1} be $y_{n+1}^{(h)}$ and the value obtained by a single step with doubled spacing $2h$ be $y_{n+1}^{(2h)}$. Then there follows (approximately)

$$Y_{n+1} - y_{n+1}^{(h)} \approx 2E_n h^{p+1}$$
$$Y_{n+1} - y_{n+1}^{(2h)} \approx 2^{p+1} E_n h^{p+1} \tag{5.42}$$

for small h. Upon eliminating E_n from these equations we obtain a Richardson extrapolation formula (see Hildebrand [13, p. 77])

$$Y_{n+1} \approx y_{n+1}^{(h)} + \frac{y_{n+1}^{(h)} - y_{n+1}^{(2h)}}{2^p - 1}. \tag{5.43}$$

Thus, if at various stages the newly calculated ordinate y_{n+1} is recomputed from y_{n-1} with a doubled spacing the truncation error in the original value is approximated by dividing the difference of the two values by $2^p - 1$. This value is 7 for third order and 15 for fourth order. Some computer programs incorporate this approximate computation automatically. Further discussion is given by Todd [12].

An evaluation of the Runge–Kutta methods is given below:

(1)   The process is convenient for automatic computation since the method is iterative.
(2)   The method is self starting unlike many finite difference methods based on the formulae of Section 5.1.
(3)   Arbitrary change of interval can be made at any stage with no appreciable complication. In finite difference methods a change of interval requires special treatment.
(4)   These methods usually *admit partial instability* (see Section 5.6);
(5)   The time of computation may be lengthy. This is a situation which can be avoided in some finite difference methods.
(6)   The estimation of local truncation error is not easily carried out except in approximation.
(7)   No method is known for determination of interval size.

Merson [32, see also 33] suggests a fourth order process which, in certain circumstances not only provides for *automatic interval size adjustment* but also an additional computation serves to develop an alternate expression to Eq. (5.43) for the error. He uses the equations

$$y_1 = y_0 + \tfrac{1}{3} hf(x_0, y_0)$$

$$y_2 = y_0 + \tfrac{1}{6} hf(x_0, y_0) + \tfrac{1}{6} hf(x_0 + \tfrac{1}{3} h, y_1)$$

$$y_3 = y_0 + \tfrac{1}{8} hf(x_0, y_0) + \tfrac{3}{8} hf(x_0 + \tfrac{1}{3} h, y_2) \qquad (5.44)$$

$$y_4 = y_0 + \tfrac{1}{2} hf(x_0, y_0) - \tfrac{3}{2} hf(x_0 + \tfrac{1}{3} h, y_2) + 2hf(x_0 + \tfrac{1}{2} h, y_3)$$

$$y_5 = y_0 + \tfrac{1}{6} hf(x_0, y_0) + \tfrac{2}{3} hf(x_0 + \tfrac{1}{2} h, y_3) + \tfrac{1}{6} hf(x_0 + h, y_4).$$

5.4 Simultaneous Equations

In our previous discussion only first order equations have been examined. The formulae are easily extended to cover the case of simultaneous first order equations and higher order equations directly. Most computers have general subroutines for simultaneous first order equations. We give two simple examples in this section.

The *standard* Runge–Kutta process applied to the *pair* of equations

$$\frac{dy}{dx} = f(x, y, z), \qquad \frac{dz}{dx} = g(x, y, z) \qquad (5.45)$$

uses the formulae

$$k_1 = f(x_n, y_n, z_n), \qquad\qquad m_1 = g(x_n, y_n, z_n),$$

$$k_2 = f(x_n + \tfrac{1}{2} h, y_n + \tfrac{1}{2} hk_1, z_n + \tfrac{1}{2} hm_1), \qquad m_2 = g \quad \text{(same as } k_2\text{)},$$

$$k_3 = f(x_n + \tfrac{1}{2} h, y_n + \tfrac{1}{2} hk_2, z_n + \tfrac{1}{2} hm_2), \qquad m_3 = g \quad \text{(same as } k_3\text{)},$$

$$k_4 = f(x_n + h, y_n + hk_3, z_n + hm_3), \qquad m_4 = g \quad \text{(same as } k_4\text{)},$$

$$y_{n+1} = y_n + \tfrac{1}{6} h[k_1 + 2k_2 + 2k_3 + k_4],$$

$$z_{n+1} = z_n + \tfrac{1}{6} h[m_1 + 2m_2 + 2m_3 + m_4]. \qquad (5.46)$$

Generalizations to three and more equations follow this same pattern.

For the second order equation

$$v'' = g(x, v, v'), \qquad v(x_0) = v_0, \qquad v'(x_0) = v_0' \qquad (5.47)$$

we find after a short calculation

$$v_{n+1} = v_n + hv_n' + \tfrac{1}{6} h^2 [k_1 + k_2 + k_3]$$
$$v_{n+1}' = v_n' + \tfrac{1}{6} h[k_1 + 2k_2 + 2k_3 + k_4]$$
$$k_1 = g(x_n, v_n, v_n')$$
$$k_2 = g(x_n + \tfrac{1}{2} h, v_n + \tfrac{1}{2} hv_n', v_n' + \tfrac{1}{2} hk_1) \qquad (5.48)$$
$$k_3 = g(x_n + \tfrac{1}{2} h, v_n + \tfrac{1}{2} hv_n' + \tfrac{1}{4} h^2 k_1, v_n' + \tfrac{1}{2} hk_2)$$
$$k_4 = g(x_n + h, v_n + hv_n' + \tfrac{1}{2} h^2 k_2, v_n' + hk_3).$$

The applications of variants of Runge–Kutta methods are legion. The minimum storage Gill modification is used by Siekmann [34] in solving the two point boundary value problem of calculating the thermal laminar boundary layer on a rotating sphere. Garg and Siekmann [35] use the Merson procedure to solve an initial value problem of three nonlinear equations arising from axisymmetric blast waves generated by a finite spherical charge. Bellman and Kalaba [36] combine the Runge-Kutta method with quasilinearization in solving $u'' = e^u$, $u(0) = u(1) = 0$ and related problems.

5.5 Multi-Step Methods†

The one step methods replace the first order differential Eq. (5.21) by a first order system of difference Eqs. (5.22). We saw that the local error committed thereby could be made as small as $O(h^5)$ (Runge–Kutta) but not smaller without much added effort. If Eq. (5.21) is replaced by difference equations of higher order the local error can be decreased to any desired order of magnitude.

The *linear multistep methods*, which concern us here, define y_{n+1} by a linear combination of y_{n-s+1}, hf_{n+1}, hf_n, ..., hf_{n-s+1}, $s = 1, 2, ..., k$ where $f_n = f(x_n, y_n)$. This "k step" method is written in the form

$$y_{n+1} = \sum_{i=0}^{k-1} a_i y_{n-i} + h \sum_{j=-1}^{k-1} b_j f_{n-j} \qquad (5.49)$$

where a_i, b_j are real constants, $k > 1$, $|a_{k-1}| + |b_{k-1}| > 0$ and $h = x_{n+1} - x_n$ is independent of n. Equation (5.49) requires the *initial* knowledge of k values $y_0, ..., y_{k-1}$ before it can be used to obtain successively

† See Todd [12].

all values y_n for $n \geq k$. Thus these methods are *not self starting*. The k initial values must be formed by some other method.

If $b_{-1} = 0$ the Eq. (5.49) is called *open* since the next approximation y_{n+1} is given explicitly in terms of the preceding values. If $b_{-1} \neq 0$ the formula is called *closed* since in this case y_{n+1} is defined implicitly by Eq. (5.49) by virtue of its implicit appearance in f_{n+1}. In the latter case Eq. (5.49) represents a nonlinear algebraic equation for y_{n+1} (or m nonlinear algebraic equations if m differential equations are present).

A theoretical treatment of these methods is given by Todd [12]. The best known of these methods are the *predictor-corrector* schemes. Under this class we find the methods of Milne [32] as modified by Hamming [38] and Keitel [39]. We discuss Hamming's method in the sequel. The Adams–Bashforth method (see Fox [5]) is also a predictor-corrector method. Hamming's treatment [40] of predictor-corrector theory and application is particularily readable.

Predictor-corrector methods are widely used because of the following *advantages*:

1. The difference between the corrected and predicted values can be used as a measure of the error made at each step and thus used to control the step size.
2. Considerable computing effort is saved because fewer derivative evaluations are required than in the Runge–Kutta method.
3. Ease of detection of machine failures.

The main *disadvantages* are:

1. Complexity of programming.
2. Lack of self starting capability.
3. The methods are more subject to catastrophic "strong" instability than are the Runge–Kutta methods.

The most generally used *linear corrector formula*, of the form given by Eq. (5.49), is

$$y_{n+1} = a_0 y_n + a_1 y_{n-1} + a_2 y_{n-2}$$
$$+ h[b_{-1} y'_{n+1} + b_0 y'_n + b_1 y'_{n-1} + b_2 y'_{n-2}]$$
$$+ E_5 \frac{h^5 y^{(5)}(\theta)}{5!}. \tag{5.50}$$

Equation (5.50) is exact for all polynomials up to and including degree four. That is, it is exact when $y = 1, x, x^2, x^3$, and x^4. Using these five conditions and taking a_1 and a_2 as parameters we obtain the relations

$$
\begin{aligned}
a_0 &= 1 - a_1 - a_2, & b_0 &= \tfrac{1}{24}(19 + 13a_1 + 8a_2), \\
a_1 &= a_1, \, a_2 = a_2, & b_1 &= \tfrac{1}{24}(-5 + 13a_1 + 32a_2), \\
b_{-1} &= \tfrac{1}{24}(9 - a_1), & b_2 &= \tfrac{1}{24}(1 - a_1 + 8a_2), \\
& & E_5 &= \tfrac{1}{6}(-19 + 11a_1 - 8a_2).
\end{aligned}
\tag{5.51}
$$

Thus five of the seven parameters are used to reduce the *truncation* error. The remaining two degrees of freedom, a_1 and a_2 are used to reduce two other types of error namely propagation error (instability) and the *amplification of round-off error*.

A most desirable *predictor*, to go with the corrector Eq. (5.50), would be one that *explicitly* uses information at the (same) preceding three points. Thus

$$
\begin{aligned}
y_{n+1} = A_0 y_n + A_1 y_{n-1} + A_2 y_{n-2} \\
+ h[B_0 y_n' + B_1 y_{n-1}' + B_2 y_{n-2}'] + \bar{E}_5 \frac{h^5 y^{(5)}}{5!}.
\end{aligned}
\tag{5.52}
$$

The coefficients are determined by asking for this result to be exact for $1, x, x^2, x^3, x^4$ or by, what is equivalent, setting $b_{-1} = 0$ in Eq. (5.51). Thus

$$
\begin{aligned}
A_0 &= -8 - A_2, & B_0 &= (17 + A_2)/3, \\
A_1 &= 9, & B_1 &= (14 + 4A_2)/3, \\
A_2 &= A_2, & B_2 &= (-1 + A_2)/3, \\
& & \bar{E}_5 &= (40 - 4A_2)/3.
\end{aligned}
\tag{5.53}
$$

The large value of A_1 means that at best we shall have a *large noise amplification* N_a in the predictor for

$$
N_a = [A_0{}^2 + A_1{}^2 + A_2{}^2]^{1/2} > 9.
$$

This large value of N_a is undesirable. To avoid it but still keep the three term predictor we ask that the formula be exact only through x^3, thereby leaving an error $O(h^4)$. This requirement leads to the coefficients

$$A_0 = 1 - A_1 - A_2, \quad B_0 = (23 + 5A_1 + 4A_2)/12,$$
$$A_1 = A_1, \quad B_1 = (-16 + 8A_1 + 16A_2)/12$$
$$A_2 = A_2, \quad B_2 = (5 - A_1 + 4A_2)/12, \quad (5.54)$$
$$\bar{E}_4 = 9 - A_1.$$

While this scheme is sometimes justified it may result, according to Hamming [40], in a waste of computation.

The predictor scheme equation (5.52) has only six parameters to use in searching for a suitable predictor. To obtain seven parameters one can add one more old value, say y_{n-3}, or one old value of the derivative y'_{n-3}. The first addition leads to *Milne type predictors*. These are

$$y_{n+1} = A_0 y_n + A_1 y_{n-1} + A_2 y_{n-2} + A_3 y_{n-3}$$
$$+ h[B_0 y_n' + B_1 y'_{n-1} + B_2 y'_{n-2}] + \bar{E}_5 \frac{h^5 y^{(5)}}{5!} \quad (5.55)$$

with

$$A_0 = -8 + A_2 + 8A_3, \quad B_0 = \frac{17 + A_2 - 9A_3}{3},$$

$$A_1 = 9 - 9A_3, \quad B_1 = \frac{14 + 4A_2 - 18A_3}{3},$$

$$A_2 = A_2, \quad B_2 = \frac{-1 + A_2 + 9A_3}{3}, \quad (5.56)$$

$$A_3 = A_3, \quad \bar{E}_5 = \frac{40 - 4A_2 + 72A_3}{3}.$$

Hamming [40] shows via an *influence function* that $A_3 \geq 0$ if the error is to be correct. If we attempt to reduce the *truncation error* $\bar{E}_5$ by setting $A_3 = 0$ we find that $A_1 = 9$ which is quite large for noise amplification. Alternatively if we try to minimize $N_a = (\sum_{i=0}^{3} A_i^2)^{1/2}$ we get

$$A_0 = -4/114, \quad A_1 = 9/114, \quad A_2 = -4/114, \quad A_3 = 113/114$$

which is very close to Milne's predictor

$$y_{n+1} = y_{n-3} + \frac{4h}{3} (2y_n' - y'_{n-1} + 2y'_{n-2}) + \frac{14}{45} h^5 y^{(5)}(\theta)$$

with $A_0 = A_1 = A_2 = 0, A_3 = 1$.

If instead of the additional term y_{n-3} we add y'_{n-3} then we obtain *Adams–Bashforth type predictors* of the form

$$
\begin{aligned}
y_{n+1} = A_0 y_n &+ A_1 y_{n-1} + A_2 y_{n-2} \\
&+ h[B_0 y'_n + B_1 y'_{n-1} + B_2 y'_{n-2} + B_3 y'_{n-3}] + \bar{E}_5 \frac{h^5 y^{(5)}}{5!}
\end{aligned} \tag{5.57}
$$

with

$$
\begin{aligned}
A_0 &= 1 - A_1 - A_2, & B_1 &= (-59 + 19A_1 + 32A_2)/24, \\
A_1 &= A_1, & B_2 &= (37 - 5A_1 + 8A_2)/24, \\
A_2 &= A_2, & B_3 &= (A_1 - 9)/24, \\
B_0 &= (55 + 9A_1 + 8A_2)/24, & \bar{E}_5 &= (251 - 19A_1 - 8A_2)/6.
\end{aligned} \tag{5.58}
$$

Some typical Adams–Bashforth predictors are given in Table 5.2.

TABLE 5.2

	Adams–Bashforth	High Accuracy	"3/8"	"1/3"	"1/2"	"2/3"
A_0	1	0	0	1/3	1/2	0
A_1	0	1	0	1/3	1/2	2/3
A_2	0	0	1	1/3	0	1/3
B_0	55/24	8/3	21/8	91/36	119/48	191/72
B_1	−59/24	−5/3	−9/8	−63/36	−99/48	−107/72
B_2	37/24	4/3	15/8	57/36	69/48	109/72
B_3	−9/24	−1/3	−3/8	−13/36	−17/48	−25/72
$\bar{E}_5$	251/6	116/3	243/6	121/3	161/4	707/18
$\bar{E}_5 - E_5{}^a$	270/6	240/6	270/6	260/6	255/6	250/6

[a] Computed for the case $a_i = A_i$ $(i = 0, 1, 2)$.

5.6 Choice of Predictor-Corrector Method

During the design of a method, the choice of corrector and predictor involves balancing of three somewhat inconsistent requirements:

(a) Large stability margin.
(b) Small truncation error.
(c) Round off error control.

Truncation error is measured by

$$\frac{E_5 h^5 y^{(5)}(\theta)}{5!};$$ (5.59)

noise amplification is measured by

$$N_C = (1 + a_1 + 2a_2)^{-1}$$ (5.60)

and to control the uncorrelated noise it is well to keep

$$N_a = \left(\sum_{i=0}^{2} a_i^2 \right)^{1/2}$$ (5.61)

reasonably small. If round off error is not to grow then $|N_C| \le 1$ and the *smaller the better*. For positive $1 + a_1 + 2a_2$ the condition for an isolated

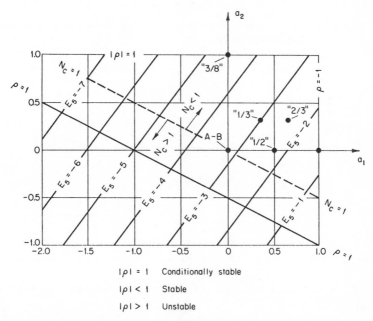

Fig. 5-1. Stability domain in (a_1, a_2) plane showing noise amplification N_C. From Hamming [40].

error not to grow is that (a_1, a_2) lie above the line $a_2 = -a_1/2$, i.e., $N_C = 1$.

The role of the corrector is the *most important* one so we discuss it first. Figure 5-1, adapted from Hamming [40] by permission, shows the domain from which we may select the coefficients a_1, a_2 for our corrector. Figure 5-2 shows the stability measure. If $A = \partial f/\partial y$ is positive in the entire interval of integration then Milne's corrector $(a_1 = 1, a_2 = 0)$ is an excellent choice. However, if $A = \partial f/\partial y$ can *be negative* then Milne's corrector is *unstable*. As a consequence we need to decide what compromise should be made between instability measured by $Ah = (\partial f/\partial y) h$ and truncation measured by $E_5 h^5 y^{(5)}(\theta)/5!$. While no explicit relation between these terms exists they tend to be large or small together.

We shall use Fig. 5-1 and 5-2 to discuss the choice of a corrector (really

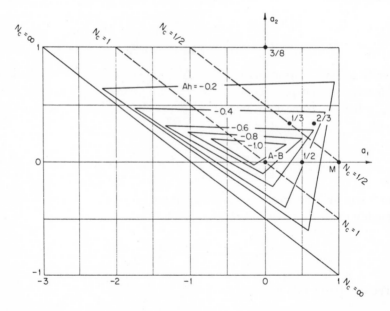

Fig. 5-2. Stability domain for predictor-corrector methods. From Hamming [40].

the choice of a_1, a_2). An arbitrary choice of $Ah = -0.4$ (and thus of h, given A) provides a safe amount of instability protection for most equations. To dimmish round-off troubles (i.e., to make N_C as small as possible) suggests moving to the upper right-hand corner of the $Ah = -0.4$ region.

Thus a choice of "2/3", where $a_1 = 2/3$, $a_2 = 1/3$ is dictated. From Table 5.1 we see that this is a mixture of 2/3 of Milne with 1/3 of the "3/8" rule.

Note that for problems which have a large negative $\partial f/\partial y$ a more stable method should be chosen. To achieve modest accuracy with more stability one immediate suggestion is to move closer to the center of gravity of the triangular stability region while staying above the line $N_C = 1$.

In selecting a predictor there is a tendency to match the a's of the corrector with the A's of the predictor. This eliminates Milne type predictors from consideration and leaves the Adams–Bashforth type which we adopt for the remainder of our discussion.

For clarity we now write p_{n+1} or C_{n+1} for the respective value calculated by the predictor or corrector formula. With matching a's and A's we find that

$$p_{n+1} - C_{n+1} = h[B_0 y_n' + B_1 y_{n-1}' + B_2 y_{n-2}' + B_3 y_{n-3}'$$
$$- (b_{-1} p_{n+1}' + b_0 y_n' + b_1 y_{n-1}' + b_2 y_{n-2}')] \qquad (5.62)$$

which acts as a measure of the fifth derivative. It is a common computational observation that $p_n - C_n$ is sensibly constant from step to step. Since p_{n+1} has the truncation error $\bar{E}_5 h^5 y^{(5)}/5!$ we modify the predictor step with

$$\frac{\bar{E}_5}{\bar{E}_5 - E_5}(p_n - C_n) \approx \frac{\bar{E}_5 h^5 y^{(5)}}{5!} \qquad (5.63)$$

to "mop up" the error (in the first step take $p_n - C_n = 0$). We similarly modify the corrector with

$$\frac{E_5 h^5 y^{(5)}}{5!} \approx \frac{E_5}{\bar{E}_5 - E_5}(p_{n+1} - C_{n+1}). \qquad (5.64)$$

The two thirds rule now has the following procedure:

Predict:

$$p_{n+1} = \frac{2}{3} y_{n-1} + \frac{1}{3} y_{n-2} + \frac{h}{72} [191 y_n' - 107 y_{n-1}' + 109 y_{n-2}' - 25 y_{n-3}']$$

$$+ \frac{707}{2160} h^5 y^{(5)}.$$

Modify:

$$m_{n+1} = p_{n+1} - \frac{707}{2160} h^5 y^{(5)} \approx p_{n+1} - \frac{707}{750}(p_n - C_n).$$

Correct:

$$C_{n+1} = \frac{2}{3} y_{n-1} + \frac{1}{3} y_{n-2} + \frac{h}{72}[25m'_{n+1} + 91y_n' + 43y'_{n-1} + 9y'_{n-2}]$$

$$- \frac{43}{2160} h^5 y^{(5)}.$$

Modify:

$$y_{n+1} = C_{n+1} + \frac{43}{2160} h^5 y^{(5)} = C_{n+1} + \frac{43}{750}(p_{n+1} - C_{n+1})$$

$$N_C = 3/5, \qquad N_a = 0.75$$

The mopping up procedure has the effect of upgrading the method to sixth order with little added computation. The reader will find it instructive to compare the 2/3 rule with the 1/2 rule. The latter is an average of the Milne and Adams–Bashforth procedures.

5.7 Boundary Value Problems

Our primary concern in the preceding material has been with initial value problems. Problems of *steady state* or *equilibrium* lead to boundary value problems in which the auxiliary conditions are shared between two or more points. The two point problems are by far the most numerous so our discussion will be limited to these. Fox [4] devotes an entire monograph to this subject. Our treatment must of necessity be very limited.

Theoretically we can solve all boundary value problems by initial value processes. In Chapter 3 a method was demonstrated whereby large classes of boundary value problems were converted to initial value problems. More generally, initial value systems are formed from the boundary value problems by adding sufficient guessed conditions at one point. These conditions are adjusted by some algorithm (repetitive operation on the analog) until the required relations are satisfied at the other point. With this approach stability problems occur occasionally and strictly boundary value methods are needed.

Initial value methods using superposition are applicable to linear systems but the approach is much more tentative for nonlinear systems. To fix the ideas and at the same time extend the process to simultaneous equations we consider the nonlinear system

$$
\begin{aligned}
y'' + yy' + yz^2 &= a(x) \\
z'' - z^3 y' + z \sin y &= b(x)
\end{aligned}
\tag{5.65}
$$

with boundary conditions

$$
y(x_0) = y_0, \qquad z(x_0) = z_0, \qquad y(x_n) = y_n, \qquad z(x_n) = z_n. \tag{5.66}
$$

For such complicated systems the Runge–Kutta method is easily applied. Thus we set $y' = u$, $z' = v$ and obtain the four first order equations

$$
\begin{aligned}
y' &= u \\
u' &= a(x) - yz^2 - yu \\
z' &= v \\
v' &= b(x) + z^3 u - z \sin y
\end{aligned}
\tag{5.67}
$$

with y_0, y_n, z_0, and z_n as given values.

To begin an initial value or step by step method beginning at x_0 we must choose two parameters, say u_0 and v_0, and subsequently adjust them to satisfy the conditions y_n and z_n at the end point x_n. That is to say

$$
y_n = y_n(u_0, v_0), \qquad z_n = z_n(u_0, v_0) \tag{5.68}
$$

and the development of the process is based upon these nonlinear equations.

For given initial values u_0, v_0 at x_0 we can calculate successive corrections $\Delta u_k = u_{k+1} - u_k$, $\Delta v_k = v_{k+1} - v_k$ from the generalized Newton–Raphson linear equations

$$
\begin{aligned}
y_n(u_k, v_k) - y_n = \Delta y_n &= \frac{\partial y_n}{\partial u_k} \Delta u_k + \frac{\partial y_n}{\partial v_k} \Delta v_k \\
z_n(u_k, v_k) - z_n = \Delta z_n &= \frac{\partial z_n}{\partial u_k} \Delta u_k + \frac{\partial z_n}{\partial v_k} \Delta v_k
\end{aligned}
\tag{5.69}
$$

where y_n and z_n represent the prescribed values at x_n and $y_n(u_k, v_k)$ the value obtained from the initial value calculation with u_k, v_k. The calculation requires knowledge of the partial derivatives of y_n and z_n with respect to u_k and v_k, evaluated at x_n. These partial derivatives can be accurately evaluated by forming differential systems from the original Eqs. (5.67) and then by integrating these step by step from x_0 to x_n using the proper initial values.

Let us set $p = \partial y/\partial u_k$, $q = \partial y/\partial v_k$, $r = \partial z/\partial u_k$, and $s = \partial z/\partial v_k$. From Eqs. (5.67) we obtain the following eight equations by differentiating them first with respect to u_k and with respect to v_k:

$$p' - (\partial u/\partial u_k) = 0$$
$$(\partial u/\partial u_k)' + pu + y(\partial u/\partial u_k) + pz^2 + 2yzr = 0$$
$$r' - (\partial v/\partial u_k) = 0$$
$$(\partial v/\partial u_k)' - z^3(\partial u/\partial u_k) - 3z^2 ur + r \sin y + zp \cos y = 0$$
$$q' - (\partial u/\partial v_k) = 0 \tag{5.70}$$
$$(\partial u/\partial v_k)' + qu + y(\partial u/\partial v_k) + qz^2 + 2yzs = 0$$
$$s' - (\partial v/\partial v_k) = 0$$
$$(\partial v/\partial v_k)' - z^3(\partial u/\partial v_k) - 3z^2 us + s \sin y + zq \cos y = 0$$

with the initial conditions

$$p = r = \partial v/\partial u_k = q = s = \partial u/\partial v_k = 0, \quad \text{at} \quad x = x_0,$$

$$\frac{\partial v}{\partial v_k} = \frac{\partial u}{\partial u_k} = 1 \qquad \text{at} \quad x = x_0. \tag{5.71}$$

When this system is integrated from x_0 to x_n the values of p, q, r, s at x_n are used in Eq. (5.69). This method requires m times as much work for m equations as in the original approximation but the convergence is ultimately *quadratic*.

An alternate of this method which has slower convergence but a smaller amount of work is given by Warner [41]. For linear problems a boundary value method is sometimes used (see Fox [4]) but the appearance of nonlinear *algebraic* equations for nonlinear equations makes this method less attractive here.

A simple iterative method employing the Gill modification of the Runge–Kutta method was employed by Siekmann [34] to calculate the velocity profile and temperature in the laminar boundary layer on a rotating sphere. An analytic continuation method coupled with the Newton system, described above, was used by Lee and Ames [42] in determining power law flow about a wedge.

5.8 Components of an Electronic Analog Computer

We now turn our attention to computation by means of electronic analog computers. These general purpose computers have evolved to the state where today a sufficient number of computer elements can be interconnected to perform, with limited accuracy ($\sim 10^{-4}$), many of the computations required in technical and scientific problems. These basic elements are constructed from *electronic amplifiers* modified by *resistance* and *capacitance* elements. Due to great bulk inductances are omitted thereby requiring the inclusion of electronic amplifiers. It is very instructive to investigate the internal design of the various elements and to see how the accuracy of 10^{-4} to 10^{-5} is attained. We shall, however, leave the design to the literature and only briefly discuss the elements.

Most analog computers use *voltage* as the *dependent* and *time* as the *independent* machine variables. These then must be the *analogs* of the physical variables in the original problem. The choice of constants of proportionality, relating problem to machine variables, is called *scaling* Determination of proper scale factors (they are not unique) is an important step in analog programming.

The analog computer is a collection of electronic circuits of a few types which can be interconnected to form the electronic analog of the problem. Consequently, we shall study the building blocks individually and then consider the procedures used to interconnect them. These basic elements can be subdivided into two classes; namely *linear* and *nonlinear* elements.

The basic linear elements the *attenuators* (potentiometers), *summers*, *integrators*, and *high gain amplifiers* are shown, together with their functions in Fig. 5-3. The fundamental circuit from which the summer and integrator are formed is the operational amplifier (*OA*) whose block diagram is shown in Fig. 5-4.

The OA is essentially a very high gain ($G \approx 10^4 - 10^8$)† dc amplifier. The

---

† The higher the gain the more expensive the equipment.

Z's represent input impedance (Z_i) and feedback impedance Z_f. There is essentially no current flow from the grid into the amplifier so the current

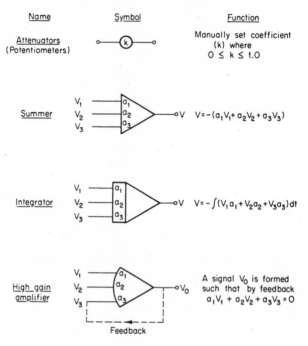

Fig. 5-3. Linear computing elements.

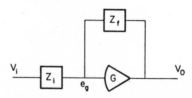

Fig. 5-4. Operational amplifier block diagram.

through Z_i must be that flowing through Z_f. Thus

$$\frac{V_i - e_g}{Z_i} + \frac{V_0 - e_g}{Z_f} = 0. \qquad (5.72)$$

But $V_0 = Ge_g$ so

$$\left(V_i - \frac{V_0}{G}\right)Z_f = -\left(V_0 - \frac{V_0}{G}\right)Z_i$$

or

$$V_0 = -V_i \frac{Z_f}{Z_i - \frac{1}{G}(Z_i + Z_0)} \tag{5.73}$$

and since $G \gg 1$ we have to a *high order* of *approximation*

$$V_0 = -V_i \frac{Z_f}{Z_i}. \tag{5.74}$$

Similarly, it can be shown that amplifiers with multiple inputs have an output

$$V_0 = -\left[V_1 \frac{Z_f}{Z_1} + V_2 \frac{Z_f}{Z_2} + V_3 \frac{Z_f}{Z_3}\right].$$

The ratio of Z_f/Z_i is called the "*gain*" of that particular signal channel. (*This is not the gain of the high gain amplifier.*)

For *summers* (summing amplifiers), the Z's are precision resistors so that $Z_f/Z_i = R_f/R_i = A$. Usually R_f is chosen as 1 megohm, while R_i is 0.1, 0.2, or 1 megohm allowing a choice of gain 10, 5, or 1. Other ratios can be used.

For *integrators*, the feedback impedance Z_f is a capacitor C and *all* the Z_i are resistors R_i. Thus we find

$$\frac{Z_f}{Z_i} = \frac{1}{R_iC}\int dt$$

so that

$$V_0 = -\frac{1}{R_iC}\int V_i\, dt \tag{5.75}$$

and for multiple inputs

$$V_0 = -\int\left(\sum_{i=1}^{n} \frac{V_i}{R_iC}\right)dt.$$

The feedback capacitor is usually 1 microfared (1 μf) or 0.1 μf. Thus with $R_i = 0.1$, 0.2, and 1.0 the integrator gains become 10, 5, and 1 volts per second with 1 μf and 100, 50, and 10 with 0.1 μf. So much for the linear elements.

There are three basic nonlinear computing elements, *multipliers, function generators*, and *discontinuous elements*. The symbol and function of some of these elements is shown in Fig. 5-5. A greater variety than those shown

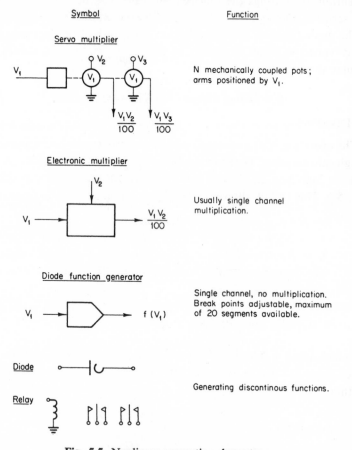

Fig. 5-5. Nonlinear computing elements.

is available (see Johnson [14], Korn and Korn [15], Karplus and Soroka [16], Warfield [17], Scott [18], and Jackson [19].) There are two types of multipliers, the electronic and the servo multipliers. The main difference is the number of available channels for multiplication. The servo device usually has more channels than the electronic multiplier. Other function generators are available but we shall discuss only the diode function generator. This device approximates a function (whether analytic or an empirical curve) by a series of straight line segments. By means of simple potentiometer settings the function generator can be made to reproduce this series of line segments when driven by the input. The main advantage of the diode function generators is that the length of the segments can be easily varied in order to obtain better approximations to curves of ir-regular shape.

5.9 Analog Circuits for Simple Operations

A. DIVISION†

The operation N/D is accomplished by a combination of a multiplier and a high gain amplifier. The equation $N/D = V_0$ is transformed to $N - DV_0 = 0$ and the high gain amplifier is used to maintain this identity as shown in Fig. 5-6. This circuit allows the high gain amplifier to provide an output

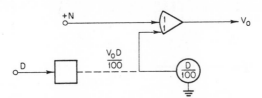

Fig. 5-6. Division circuit.

V_0 which, when multiplied by $D/100$, will balance in sign and amplitude the other input N—that is, $N + V_0D/100 = 0$, $V_0 = -(N/D)\cdot 100$. Alternate methods are available.

B. SQUARE ROOT

A square root circuit can be constructed with a function generator or in the manner below. If given x, $V_0 = \sqrt{x}$ is desired we rearrange the equation as $x - V_0^2 = 0$ and mechanize as in Fig. 5-7. A scale factor of 10 appears for if x ranges from 0 to 100 then $\sqrt{x}$ ranges from 0 to 10.

---

† In the sequel we assume our computer has a ±100 volt range.

C. Higher Powers

The second and higher powers are obtained by successive multiplications or by function generators. A typical repeated multiplication circuit using a servo multiplier to obtain x^2, x^3, and x^4 is shown in Fig. 5-8.

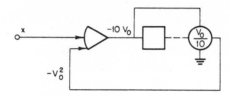

Fig. 5-7. Square root circuit.

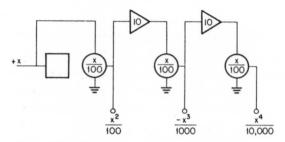

Fig. 5-8. Higher powers using a servo-multiplier.

D. Function Generator

Empirical functions are usually simulated with any of the variety of special function generators. If analytic expressions of the independent variable, say t, are needed one may take the alternate of generating the function via its defining differential equation. Thus if $y = \sin \omega t$ is required we solve its differential equation

$$\frac{d^2y}{dt^2} + \omega^2 y = 0, \qquad y(0) = 0, \qquad y'(0) = \omega. \tag{5.76}$$

E. Transcendental Functions of y

If one of the coefficients is a function of the dependent variable y, the auxiliary equation used to generate this coefficient (in the spirit of part D) has as its independent variable the dependent variable of the differential equation. Thus the given and auxiliary differential equations have different independent variables. This is a matter of considerable concern because the single independent variable of the analog, real time, cannot simultaneously represent both. To avoid this difficulty we use the relation

$$\int f(y)\, dy = \int f(y)\left(\frac{dy}{dt}\right) dt. \tag{5.77}$$

Thus integration with respect to the dependent variable y can be performed, if both dy/dt and a multiplier are available, by an integrator whose integration variable is t. The first derivative is normally available in such problems. An example problem will illustrate this procedure in Section 5.11.

5.10 Scaling Procedures and Initial Conditions

Analog computer simulation of physical problems requires two separate scaling procedures, one for signal amplitude (dependent variables) and the other for time (independent variable). This results because of the necessity of defining the analogy between the computer variables (time and voltage) and the problem variables (such as distance, angle, pressure, temperature, velocity, concentration, etc). Certainly, in some cases, these scale factors may be unity—that is, the numerical value of the computer voltage equals the numerical value of the problem variable. This procedure is related to the nondimensional procedures that we customarily follow in any computational problem. The ability to change time scale is one of the major advantages of the analog computer in that a physical problem can be simulated on an expanded or contracted time scale. However, one does not have complete flexibility because of the following physical and electronic limitations:

(a) Integration errors accumulate in long runs, that is, if e is an error in a signal E_i

$$\int (E_i + e)\, dt = \int E_i\, dt + et. \tag{5.78}$$

(b) Servo-mechanisms, often used for multiplying, operate best at low frequencies (< 5 cycles/sec).

(c) At high computing speeds amplifier phase shift is accentuated.
(d) Inertial effects restrict the range of operation of recording equipment ($X - Y$ recorders at 1 rad/sec).

A. TIME SCALING

The integrators are the elements which usually determine the speed of computation. The basic time unit is always a second. When one volt is connected to an integrator with gain 1 the result will be -1.0 volt/sec on the output. This results since the integrator time constant is the product of the input resistor R and the feedback capacitor C; $RC = 10^6 \times 10^{-6}$ ((volts/coul/sec) $\cdot$ (coul/volt)) $= 1$ sec. If a 0.1 microfarad capacitor is used the result will be -10 volts/sec output—that is to say the, integrator will operate ten times faster.

The simplest method of time† scaling is as follows:

(a) Reduce the problem "time" units to seconds.
(b) Suppose we desire to change the time scale by a factor a, $a > 0$. Let τ be the computer time and t be the problem independent variable. If we set $\tau = at$, that is $t = \tau/a$ and $a > 1$, the solution is slowed by a factor a. If $a < 1$ the problem is speeded by a factor $1/a$. This is most easily seen by comparing the ranges

$$
\begin{array}{ll}
\text{real time:} & 0 < t < t_0 \\
\text{computer time:} & 0 < \tau < at_0.
\end{array}
\tag{5.79}
$$

Thus if $a > 1$ the computer takes longer than real time and if $a < 1$ the computer takes less than real time to accomplish the solution. This is one of the highly desirable features of electronic analogs.

To time scale we make the simple change of variable

$$
t = \tau/a, \qquad \frac{d}{dt} = \frac{d}{d(\tau/a)} = a\,\frac{d}{d\tau}
\tag{5.80}
$$

and more generally

$$
\frac{d^n}{dt^n} = a^n\,\frac{d^n}{d\tau^n}.
\tag{5.81}
$$

---

† The independent variable need not be time. In these cases we identify the real problem units with the analog time unit in some way.

How should we choose a? Normally we would like the computer runs to be of short duration (a rough rule of thumb is 10 sec to 1 min) but not too short. This minimizes the effects of integrator error drift.

An approximate idea of *angular frequencies* in oscillatory problems or the *time constants* in nonoscillatory problems is often useful in estimating the value of a. In oscillatory problems one can obtain an approximate angular frequency by neglecting coupling and damping terms and then calculate the undamped, uncoupled natural angular frequency. For example consider

$$10y'' + yz + 3y' + 100y = \sin t,$$
$$6z'' + 2yz + \tfrac{1}{2} z' + 600 z = 0. \tag{5.82}$$

We depress the coupling terms yz and $2yz$ and the damping terms $3y'$ and $\tfrac{1}{2}z'$. The resulting equations have angular frequencies $\sqrt{10}$ and $\sqrt{100} = 10$.

In nonoscillatory problems we are interested in the *time constant η* which is the time in which the dependent variable reaches $1/e$ of its initial value. We usually continue the computation to approximately 4 times η at which point 1.8% of the initial value remains.

To estimate the time constant η we go through the *neglect of terms* used in the previous procedure. Thus suppose we couple to Eqs. (5.82)

$$ax' + bx + yx = z. \tag{5.83}$$

Neglecting yz and z we have

$$ax' + bx = 0$$

whose solution is

$$x(t) = x(0) e^{-(b/a)t} \tag{5.84}$$

Now $x = x(0) e^{-1}$ when $t = a/b$. a/b is an approximation to the time constant of the equation.

Of course our estimates of the frequency and time constants may be inaccurate for many reasons. Thus when the computer is operating we may wish to speed or slow the computation time. To speed the solution by a factor of α multiply *all* input signals to *all* integrations by α. For $\alpha = 2, 5,$ or 10 this can probably be accomplished on the problem plugboard. Otherwise potentiometors must be used.

B. Amplitude Scaling

In amplitude scaling we keep the following points in mind:

(a)   Voltage levels are limited to ± 100 volts. Avoid small and very large (near 100) maximum voltages into or out of any amplifier. A rule of thumb, which allows a margin of error, is to keep all peak voltages (inputs and outputs) in the ± 50 range. With this safety factor even poor estimates of maximum amplitudes are acceptable.

(b)   Use simplicity, keep the relation between the physical and the analog system as simple as possible.

(c)   Estimate the range of the problem variables as, say, y_{max} and y_{min}. If y is not symmetric about the 0 axis use y_{min}, if it is symmetric take y_{min} as zero.

For decreasing nonoscillatory problems with governing first order differential equations we can often estimate $y_{max} = y(0)$. The quantity y_{min} is taken as zero. For damped oscillatory problems with second order governing equations we can often estimate y_{max} by means of energy totals in the system at $t = 0$.

(d)   Introduce a dimensionless variable, such as

$$\bar{y} = A\left[\frac{y - y_{min}}{y_{max} - y_{min}}\right].\tag{5.85}$$

The range of the bracketed expression is from -1 to $+1$. Thus we can set $A = 50$, say, and be well within our ± 100 range for $\bar{y}$. Other values of A may be used. The equation is converted to dimensionless form and is now an equation in $\bar{y}$.

(e)   *Do not forget* to convert boundary conditions and initial conditions into the new dependent variables and new time variable τ.

C. Initial Conditions

Initial condition voltage is applied *to all and only to* integrator circuits by means of some such circuit as that shown in Fig. 5-9a. Relays 1, 2, 3 are connected to the operate-reset switch in such a manner that in *reset* relay 1 grounds out the normal input signals, relay 2 connects the resistor across the capacitor, and relay 3 applies the initial voltage. In the reset position, with initial voltage e_i, the circuit operates as a summer so the output is

$e_0 = -e_i$. The initial condition is shown on the computer diagram in the manner of Fig. 5-9b.

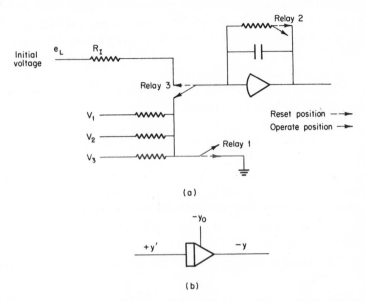

(a)

(b)

Fig. 5-9. (a) Initial condition circuit. (b) Initial condition representation on computer diagram.

5.11 Examples

(a)   As a first example we program the Siekmann equation (3.133a)

$$w''' + 2ww'' + (1 - w'^2) = 0 \qquad (5.86)$$

subject to the initial conditions

$$w(0) = w'(0) = 0, \qquad w''(0) = \lambda. \qquad (5.87)$$

In this first example we omit the scaling. For programming purposes it is convenient to rearrange the equation so that the highest order derivative is expressed in terms of the remaining quantities. Thus

$$\frac{d^3w}{dt^3} = -2w\frac{d^2w}{dt^2} - 1 + \left(\frac{dw}{dt}\right)^2. \qquad (5.88)$$

We now assume that all the terms on the righthand side of Eq. (5.88) are available and apply them, with the proper algebraic sign, to integrator 1 of Fig. 5-10. This is the same as applying d^3w/dt^3 to integrator 1. Upon

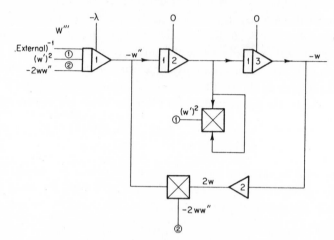

Fig. 5-10. Unscaled analog program for the Siekmann equation (3.133a).

integrating we have $-d^2w/dt^2$, together with the initial condition $-\lambda$ for that term. The remainder of the computation is easily followed by examining the course of the arrows.

(b) As a second example we present the scaled analog program for calculating the angular position $\theta = \theta(t)$ of a pendulum. The equation of motion is

$$\theta'' + 16\sin\theta = 0, \qquad \theta(0) = 1, \qquad \theta'(0) = 0. \qquad (5.89)$$

During this computation it is necessary to produce the function $y = \sin\theta$ which is a solution of the auxiliary equation $d^2y/d\theta^2 + y = 0, y[\theta(0)]$, $y'[\theta(0)]$. In solving this auxiliary equation the remarks of Section 5.9e apply—that is, Eq. (5.77) becomes

$$\int f(\theta)\,d\theta = \int f(\theta)\left(\frac{d\theta}{dt}\right) dt$$

and both θ and $d\theta/dt$ will be available from the solution of the differential equation.

Magnitude scaling is performed by considering the linearized equation

$$\theta'' + 16\theta = 0, \qquad \theta(0) = 1, \qquad \theta'(0) = 0 \qquad (5.90)$$

with solution $\theta = \cos 4t$. No time scaling appears necessary—therefore our program will be a real time solution. The maximum value of θ, $\theta_{max} = 1$, $\theta'_{max} = 4$, $\theta''_{max} = 16$. Replacement of θ by $\sin \theta$ in the simplified Eq. (5.90) would yield a smaller value of θ and hence of θ', since $|\sin \theta| \leq |\theta|$. The assumed maximum values are therefore *conservative* estimates. These remarks help us to choose the normalized quantities θ, $\theta'/4$ and $\theta''/16$. In these variables the differential Eq. (5.89) becomes†

$$\frac{\theta''}{16} + \sin \theta = 0, \qquad \theta(0) = 1, \qquad \frac{\theta'(0)}{4} = 0. \qquad (5.91)$$

Care must be exercised in writing the initial conditions for the auxiliary equation generating $\sin \theta$. These are not for *zero* θ but for *initial* θ. Since $\theta(0) = 1$, the required values are

$$y[\theta(0)] = \sin 1 = 0.8415 \qquad \text{and} \qquad y'[\theta(0)] = \cos 1 = 0.5403.$$

The scaled analog program is presented in Fig. 5-11.

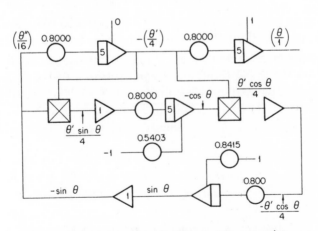

Fig. 5-11. Analog program for the pendulum equation.

(c) As a last example we describe a reaction kinetics problem that was

---

† We purposely omit a magnification factor to bring this to ± 50 volts.

programmed and solved on a medium sized analog. Many such problems are solved each year. The interesting and challenging portions of this problem arise from the continuous feed V_C of component A with constant concentration $\bar{C}_A$ and the intermittent periodic constant feed of the second component B with concentration $\bar{C}_B$. These react with kinetics

$$A + B \xrightarrow{K} C + 2D \qquad (5.92)$$

in a constant volume V constant temperature reactor. The reactor operates in cycles:

Cycle One;

Equations:

$$\frac{dC_A}{dt} = \alpha[\bar{C}_A - C_A] - KC_AC_B \qquad (5.93)$$

$$\frac{dC_B}{dt} = - KC_AC_B - \alpha C_B \qquad (5.94)$$

$$\alpha = V_C/V$$

Initial conditions:

$$C_A(0) = 0, \qquad C_B(0) = \beta\,\bar{C}_B/(1 + \beta),$$

where β is a dimensionless measure of the input volume of component B.

Equations (5.93) and (5.94) are integrated from $t = 0$ to a time t_1 at which the average overflow concentration of B is $\gamma\%$ of the initial concentration. This determines the time for the next cycle to begin—that is, for the introduction of the next slug of component B. Thus we wish to determine t_1 so that

$$\frac{\displaystyle\int_0^t C_B(t)\,dt}{t} = \gamma\,C_B(0) = \gamma\beta\,\bar{C}_B/(1 + \beta). \qquad (5.95)$$

At $t = t_1$ cycle two begins. Its equations are Eqs. (5.93) and (5.94) and its initial conditions are

$$C_{A2}(t_1) = C_{A1}(t_1)/(1 + \beta)$$
$$C_{B2}(t_1) = [C_{B1}(t_1) + \beta\,\bar{C}_B]/(1 + \beta)$$

and so forth. A sample calculation is shown in Fig. 5-12.

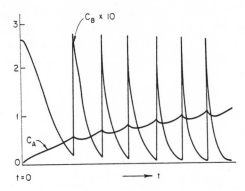

Fig. 5-12. Analog computer solutions for a batch-continuous reactor.

REFERENCES

1. Ames, W. F., and Robinson, J. A., The Model T Computer. *J. Eng. Educ.* **54,** 338 (1964).
2. Bennett, A. A., Milne, W. E., and Bateman, H., "Numerical Integration of Differential Equations," Dover, New York, 1956.
3. Collatz, L., "The Numerical Treatment of Differential Equations," Springer, Berlin, 1960.
4. Fox, L., "Numerical Solution of Two-point Boundary Problems," Oxford Univ. Press, (Clarendon), London and New York, 1957.
5. Fox, L., ed., "Numerical Solution of Ordinary and Partial Differential Equations," Pergamon Press, Oxford, 1962.
6. Levy, H., and Baggott, E. A., "Numerical Studies in Differential Equations," Dover, New York, 1956.
7. Milne, W. E., "Numerical Solution of Differential Equations." Wiley, New York, 1953.
8. Mikeladze, S. E., "New Methods of Integration of Differential Equations and Their Applications to Problems in the Theory of Elasticity," (in Russian). Gosudarstv. Izdat Tehn.-Teor. Lit., Moscow, 1951.
9. von Sanden, H., "Praxis der Differentialgleichungen," 4th ed. de Gruyter, Berlin, 1955.
10. Henrici, P., "Discrete Variable Methods in Ordinary Differential Equations," Wiley, New York, 1962.
11. Henrici, P., "Error Propagation for Difference Methods," Wiley, New York, 1963.
12. Todd, J., ed., "Survey of Numerical Analysis," McGraw-Hill, New York, 1962.

13. Hildebrand, F. B., "Introduction to Numerical Analysis." McGraw-Hill, New York, 1956.
14. Johnson, C. L., "Analog Computer Techniques," 2nd ed., McGraw-Hill, New York, 1963.
15. Korn, G. A., and Korn, T. M., "Electronic Analog Computers," 2nd ed. McGraw-Hill, New York, 1956.
16. Karplus, W. J., and Soroka, W. W., "Analog Methods," 2nd. ed, McGraw-Hill, New York, 1959.
17. Warfield, J. N., "Electronic Analog Computers," Prentice-Hall, Englewood Cliffs, New Jersey, 1959.
18. Scott, N. R., "Analog and Digital Computer Technology," McGraw-Hill, New York, 1960.
19. Jackson, A. S., "Analog Computation," McGraw-Hill, New York, 1960.
20. Jordan, C., "Calculus of Finite Differences," 2nd ed. Chelsea, New York, 1947.
21. Milne-Thompson, L. M., "Calculus of Finite Differences," Macmillan, New York, 1933.
22. Runge, C., *Math. Ann.* **46**, 167 (1895).
23. Kutta, W., *Z. Angew. Math. Phys.* **46**, 435 (1901).
24. Bashforth, F., and Adams, J. C., "An Attempt te Test The Theories of Capillary Action . . .", Cambridge Univ. Press, Cambridge, 1883.
25. Euler, L., "Institutiones Calculi Integralis." Oeuvres complètes, Berlin, 1913.
26. Wilson, E. M., *Quart. J. Mech. Appl. Math.* **2**, 208 (1949).
27. Gibbons, A., *Comput. J.* **3**, 108 (1960).
28. Heun, K., *Z. Angew. Math. Phys.* **45**, 435 (1900)
29. Kopal, Z., "Numerical Analysis," Wiley, New York, 1955.
30. Martin, D. W., *Comput. J.* **1**, 118 (1958).
31. Gill, S., *Proc. Cambridge Phil. Soc.* **47**, 96 (1951).
32. Merson, R. H., *Proc. Symp. Data Process. Weapons Res. Estab., Salisbury, South Australia, 1957.* (See Fox [5].)
33. Buckingham, R. A., "Numerical Methods," Pitman, New York, 1962.
34. Siekmann, J., *Z. Angew. Math. Phys.* **13**, 468 (1962).
35. Garg, S. K., and Siekmann, J., *Z. Angew. Math. Phys.* **17**, 108 (1966).
36. Bellman, R. E., and Kalaba, R. E., "Quasilinearization and Nonlinear Boundary Value Problems," Am. Elsevier, New York, 1965.
37. Milne, W. E., *Am. Math. Monthly* **33**, 455 (1926).
38. Hamming, R. W., *J. Assoc. Comput. Mach.* **6**, 37 (1959).
39. Keitel, G. H., *J. Assoc. Comput. Mach.* **3**, 212 (1956).
40. Hamming, R. W., "Numerical Methods for Scientists and Engineers," McGraw-Hill, New York, 1962.
41. Warner, F. J., *Math. Tab. Aids Comput.* **11**, 268 (1957).
42. Lee, S. Y., and Ames, W. F., *A.I.Ch.E. J.* **12**, 700 (1966).

Appendix

SIMILARITY VARIABLES BY TRANSFORMATION GROUPS

The material of the appendix is taken from Ames [1, 2].

The transformation group procedure appears to have been first applied by Birkhoff [3] and a general theory has evolved from the work of Morgan [4] based on theorems of Michal [5]. Morgan's method implies that the search for similarity solutions of a system of partial differential equations is equivalent to the determination of the invariant solutions of these equations under a particular one (or more) parameter group of transformations. A set of similarity variables (invariants of the group) is obtained by solving a set of simultaneous algebraic equations. This method is simple and straightforward in its application. It is easily carried through for complex systems without the cumbersome operations of the other methods.

The basic theory of the group method is summarized in the following three theorems.

Let $\sum$ be a set of N partial differential equations given by

$$\psi_j(x_i, y_j) = 0 \qquad\qquad (A.1)$$

where $x_i \, (i = 1, 2, \ldots, M)$ and $y_j \, (j = 1, 2, \ldots, N)$ are the independent and dependent variables, respectively.

Theorem I

A.   Let G_1 be the transformation group, with nonzero real parameter a,

251

whose elements are

$$\bar{x}_i = a^{\alpha_i} x_i; \qquad \bar{y}_j = a^{\lambda_j} y_j. \qquad (A.2)$$

The α_i and λ_j are to be determined so that set $\sum$ is absolutely constant conformally invariant under group G_1. A set of functions $H_j(x_i)$ is said to be "conformally invariant" under a one-parameter (a) group of the form of Eq. (A.2) if $H_j(x_i) = F_j(\bar{x}_i; a)H_j(\bar{x}_i)$, $i = 1, 2, \ldots, M$; $j = 1, 2, \ldots, N$, where the $H_j(\bar{x}_i)$ are exactly the same functions of the $\bar{x}_i$ as the H_j are of the x_i. These functions are said to be "constant conformally invariant" if the F_j are independent of the $\bar{x}_i$ and "absolutely constant conformally invariant" if $F_j = 1$ for all j. The latter definition is used exclusively in this Appendix.

From this requirement one obtains a set of simultaneous equations, nontrivial solutions of which give the invariants of G_1, which are the similarity variables.

Suppose x_k is the independent variable to be eliminated. For $\alpha_k \neq 0$, the invariants of G_1 are

$$\phi_i = \frac{x_i}{x_k^{\alpha_i/\alpha_k}} \qquad (A.3)$$

$$f_j(\phi_i) = \frac{y_j(x_1, \ldots, x_M)}{x_k^{\lambda_j/\alpha_k}}$$

where $i = 1, 2, \ldots, M$ $(i \neq k)$ and $j = 1, 2, \ldots, N$. In what follows, the range of i and j is as given in Eq. (A.3) unless otherwise specified.

B. For $\alpha_k = 0$ let G_2 be the group consisting of

$$\bar{x}_k = x_k + \ln a; \qquad \bar{x}_i = a^{\alpha_i} x_i \qquad i \neq k,$$
$$\bar{y}_j = a^{\lambda_j} y_j. \qquad (A.4)$$

The corresponding invariants (similarity variables) of G_2 are

$$\phi_i = \frac{x_i}{\exp[\alpha_i x_k]}, \qquad i \neq k$$

$$f_j(\phi_i) = \frac{y_j(x_1, \ldots, x_M)}{\exp[\lambda_j x_k]}. \qquad (A.5)$$

Extension of Theorem I to eliminate two or more independent variables simultaneously and to two or more parameter groups is possible. Theorems II and III show the immediate extensions of Theorem I.

Theorem II

A. let G_1 be the group given by Eq. (A.2) and suppose x_k and x_l are to be eliminated simultaneously. If $\alpha_k = \alpha_l \neq 0$, the invariants of G_1 are

$$\phi_i = \frac{x_i}{(bx_k + dx_l)^{\alpha_i/\alpha_k}}, \qquad i \neq k, l$$

$$f_j(\phi_i) = \frac{y_j(x_1, ..., x_M)}{(bx_k + dx_l)^{\lambda_j/\alpha_k}}, \tag{A.6}$$

and b, d are arbitrary constants.

B. If $\alpha_k = \alpha_l = 0$, the use of group G_2, Eq. (A.4), gives the invariants

$$\phi_i = \frac{x_i}{\exp[(bx_k + dx_l + e)\alpha_i]}, \qquad i \neq k, l$$

$$f_j = \frac{y_j}{\exp[(bx_k + dx_l + e)\lambda_j]} \tag{A.7}$$

where b, d, and e are arbitrary constants.

The situation for a group with more than one parameter is somewhat more complicated. In illustration, we use a two-parameter group G_3 and describe the simultaneous elimination of two independent variables.

Theorem III

Let G_3 be the two-parameter group, with real positive parameters a and b, consisting of the transformations

$$\bar{x}_i = a^{\alpha_i} b^{\beta_i} x_i, \qquad i = 1, 2, ..., M, \qquad i \neq k, l,$$

$$\bar{x}_k = a^{\alpha_k} x_k, \qquad \bar{x}_l = b^{\beta_l} x_l, \tag{A.8}$$

$$\bar{y}_j = a^{\gamma_j} b^{\delta_i} y_j, \qquad j = 1, 2, ..., N.$$

The invariants of G_3, under the requirement that system $\sum$ of partial differential equations be absolutely constant conformally invariant, are the similarity variables. Four cases arise. In all these cases $i \neq k, l$ and

$i = 1, 2, \ldots, M$, and $j = 1, 2, \ldots, N$.

A. $\alpha_k \neq 0, \beta_l \neq 0$:

$$\phi_i = \frac{x_i}{x_k^{\alpha_i/\alpha_k} x_l^{\beta_i/\beta_l}}, \qquad f_j = \frac{y_j}{x_k^{\gamma_j/\alpha_k} x_l^{\delta_j/\beta_l}}. \tag{A.9}$$

B. $\alpha_k \neq 0, \beta_l = 0$:

$$\phi_i = \frac{x_i}{x_k^{\alpha_i/\alpha_k} \exp[\beta_i x_l]}, \qquad f_j = \frac{y_j}{x_k^{\gamma_i/\alpha_k} \exp[\delta_j x_l]}. \tag{A.10}$$

C. $\alpha_k = 0, \beta_l \neq 0$:

$$\phi_i = \frac{x_i}{x_l^{\beta_i/\beta_l} \exp[\alpha_i x_k]}, \qquad f_j = \frac{y_j}{x_l^{\delta_i/\beta_l} \exp[\gamma_j x_k]}. \tag{A.11}$$

D. $\alpha_k = \beta_l = 0$:

$$\phi_i = \frac{x_i}{\exp[\alpha_i x_k + \beta_i x_l]}, \qquad f_j = \frac{y_j}{\exp[\gamma_j x_k + \delta_j x_l]}. \tag{A.12}$$

Applications of this method have been made by Ames [1, 2], Manohar [6], Lee and Ames [7], Hansen [8], and others.

REFERENCES

1. Ames, W. F., *Ind. Eng. Chem. Fundamentals* **4,** 72 (1965).
2. Ames, W. F., "Nonlinear Partial Differential Equations in Engineering," Academic Press, New York, 1965.
3. Birkhoff, G., "Hydrodynamics," 2nd ed. Princeton Univ. Press, Princeton New Jersey, 1960.
4. Morgan, A. J. A., *Quart. J. Math. Oxford Ser.* **3,** 250 (1952).
5. Michal, A. D., *Proc. Natl. Acad. Sci. U.S.* **37,** 623 (1952).
6. Manohar, R., Some similarity solutions of partial differential equations of boundary layers, Tech. Summ. Rept. 375, January 1963. Math. Res. Center, Univ. of Wisconsin, Madison, Wisconsin, 1963,
7. Lee, S. Y., and Ames, W. F., *A.I.Ch.E. J.* **12,** 700 (1966).
8. Hansen, A. G., "Similarity Analysis of Boundary Value Problems in Engineering," Prentice-Hall, Englewood Cliffs, New Jersey, 1964.

Author Index

Numbers in parentheses are reference numbers and indicate that an author's work is referred to although his name is not cited in the text. Numbers in *italics* show the page on which the complete reference is listed.

A

Abraham, W. H., 94, *132*

Abramowitz, M., 47 (19), *85*, 162, 163, *206*

Acrivos, A., 16, *22*, 184 (57), *207*

Adams, J. C., 216, *250*

Albert, A. A., 88, *132*

Alksne, A., 127, 132, *133*

Ames, W. F., 7, 8 (21), 9, 11, 13, 15, 16, 18 (50), *21*, *22*, 30, 45, 46, 61, 66, 79, *84*, *85*, 87, 94, 109, 115, 128, 130, *132*, *133*, *134*, 166, 185, *206*, 211, 235, *249*, *250*, 251, *254*

Andrews, A. H., 44 (12), *84*

Andronow, A, 6, *20*, 203, *208*

B

Baggotte, E. A., 211, *249*

Bashforth, F., 216, *250*

Bateman, H., 11, *22*, 101, *132*, 211 (2), *249*

Bautin, N. N., 170, *206*

Behn, V. C., 18 (51, 52), *22*

Bellman, R. E., 166, 170, 196, 198, 199, 200, 201, 202, *206*, *208*, 224, *250*

Bennett, A. A., 211, *249*

Benson, S. W., 8, 9, *21*, 89, *132*

Bhagavantham, S., 66, *86*

Bharucha-Reid, A. T., 95, *132*

Bickley, W. G., 7, *21*, 42, *84*, 109, *133*, 189, *207*

Birkhoff, G., 13, *22*, 251, *254*

Blodgett, K. B., 45 (16), *85*

Bogoliuboff, N. N., 7, *21*, 166, 203, *206*, 208

Bowen, J. R., 184, *207*

Boyer, R. H., 127, *134*

Bram, J., 7, *21*, 203, *208*

Bretherton, F. P., 174, *207*

Brown, O. E., 26, 59, *84*

Buckingham, R. A., 233 (33), *250*

Burgers, J. M., 11, *22*

Byrd, P. F., 51, *85*

C

Calogero, F., 199, *208*

Cap, F., 204, *208*

Carlson, A. J., 44, *84*

Carrier, G. F., 95, 125, *132*, *133*, 184, *207*

Chaikin, C. E., 203, *208*

Chaikin, S., 6, *20*

Chambré, P. L., 11, *22*, 42, *84*, 102, *133*

Chandrasekhar, S., 6, *20*, 105, *133*

Chevally, C., 66, *86*

Churchill, R. V., 95, *132*

Cole, J. D., 11, 14, *22*, 174, *207*

Collatz, L., 184, 186, 190, *207*, *208*, 211, *249*

Collings, W. Z., 18 (51, 52), *22*

Comrie, L. J., 47 (21), *85*

Crandall, S. H., 19, 20, *22*, 184, 189, *207*

Crank, J., 128, *134*

Cunningham, W. J., 7, *21*, 34, *84*, 166, 202, 203, *206*

D

Davenport, H., 205, *208*

Davies, T.V., 137, 138, 152, 165, 170, 183, 184, 203, *205*, *206*

Davis, H. T., 6, *20*, 36, 37, 42, 44, 55, *84*, 105, *133*

Dietz, J. H., 204 (81), *208*

Dressel, F. G., 64, *85*

Dunkel, O., 138 (3), *205*

Subject Index

A

Ad hoc methods, equations from, 11, 16
Airfoil surface
  equation, 5
  fluid separation, 5
Analog computation, 235–249
  basic computing elements, 235–239
  errors, 241, 242
  examples, 245–249
  scaling, 235
  variables, 235
Analytic continuation, 147, 148
Anisentropic flow of gas
  auxiliary function, 114
  equations, 114
Approximate methods, general concepts, 135, 136
Arrhenius relation, 9, 10, 101
Asymptotic matching principle, 175
Asymptotic notation, 174
Autonomous equation, 31, 152, 153

B

Bacterial growth, 19
Batch-continuous reactor, analog circuit, 248, 249
Bernoulli equation, 34
Bessel equation, 37
Biological systems, diffusion and reaction, 18, 19
Birth and death processes, 18
Blasius equation, 136
  constant of homology, 120
  conversion to integral form, 118, 119
  invariant under transformation, 120
  Taylor series solution, 141, 142
Blasius problem
  series solution, 122, 141, 142
  solution by integral methods, 161–165
  velocity profile, 162
Boundary conditions
  essential, 186

natural, 186
  quasilinearization of nonlinear, 201
Boundary layer, 174
Boundary layer flow in convergent channel, 111
  similarity variables, 111
Boundary value problem
  conversion to initial value problem, 121–127
  numerical methods, 232–235
  Runge-Kutta procedure, 233–235
Burger's equation, 11
  modified, 14

C

Calculation methods
  multi-step, 224–232
  one step, 216–224
Canonical form, for nonlinear equations, 87–94
Capillary curve equation, 45
  solution, 46, 47
Cauchy-Picard iteration, 148–150
  acceleration of convergence, 150
  convergence, 149
  equivalent difference-differential form, 149
  integral form, 149
  modified, 150–152
Chebyshev economization, of Taylor series, 142–147
Chebyshev polynomials
  approximation by, 144
  definition, 143
  economization of Taylor series, 145–147
  equal ripple property, 143
  table, 144
Chemical reaction
  constant temperature, 9
  constant volume, 9
  equations, 8–10